LIFE

Valentin Matcas, M.Ed.

Copyright © 2026 Valentin Matcas

All rights reserved.

ISBN: 9781973509363

DEDICATION

I dedicate this book to everyone eager to learn and develop continuously throughout life.

CONTENTS

1	The Living, the Dead, and the Consensual	1
2	The Essence of Life	37
3	The Source of Life	251
4	Life, Consciousness, and the Supreme	269
5	Life in all Her Forms and Realities	274
6	The Origins of Life	300
7	Model for the Organic Life	313

1 THE LIVING, THE DEAD, AND THE CONSENSUAL

You are an intelligent living human being, and in order to assure your meaning and success throughout life, you have to know everything about life everywhere, in all forms and realities. You have to learn everything about nature, society, and organic life, about the true origins of life, about creating life and about the divine, and about your own meaning in life and in the world as an intelligent living human being. These are not random ideas, but they are natural needs for higher knowledge, embedded continuously within your own intelligent human needs and meanings, as it is the case with all intelligent life.

When you are capable to find the necessary knowledge about life, meaning, society, and the world, you are truly capable to live your life at the intelligent human level. While if you cannot find it, you keep on searching, since your own higher level needs and meanings never leave you alone until you learn everything necessary in life and in the world. Otherwise, you end up living your life on lower developmental levels, addicted, in tyranny, in servitude, or only intuitively, through animal instincts. It certainly matters, because you are an intelligent living human being by nature, forced to live life in ignorance, unfulfilled, and even punished intrinsically, for your continuous lack of development.

Throughout this book, we model life in all significant details, as we study everything alive and intelligent, from the smallest cellular components to the entire human body, mind, and spirit, and to all forms of life, because everything is alive and meaningful in the world. If you want to learn more about life in all forms and realities, this book is for you.

What is so important to know about life, now grasping your continuous attention? Everything, certainly, and not only what you learn in school, about the random wild animals in the woods, about your wicked living nature, or about your tightly designated place in society. What else can there be about life, worth enough to invest all your time and effort to find it? Knowledge, all knowledge, and cognition in general, including how you can manage to know everything in a living manner.

Everything relates to cognition, while cognition is integral part of life. Because at the intelligent human level, you are continuously aspiring to detach from sweet ignorance and from unavoidable mediocrity, regardless of how strongly these pull you back, and this is how you persist to find a proper lifestyle, allowing your continuous development. The bad part is that you always lack the meaningful accurate knowledge helping you develop, since it is nowhere in the mainstream or in the alternative. You are strongly motivated to learn and outperform yourself, if you really seek to unlock the genuine understanding of life, the one that science offers only through trivial information, below your expectation. Yet how can you ever identify these, if you cannot understand life and your own meaning in life and in the world? What is life, and what is your meaning in life?

Let us study briefly the official definition of life, as it is offered by science. Yet it is not exactly a definition for life, since science does not have one. Science offers a succession of statements about life, empirical and mostly trivial, but science never actually defines life.

Life

1. Life is a characteristic distinguishing the living from the nonliving.
2. The smallest contiguous unit of life is called an organism.
3. Organisms are composed of one or more cells.
4. Organisms undergo metabolism, maintain homeostasis, grow, respond to stimuli, reproduce, and adapt to their environment in successive generations.

This is the official 'definition' of life, taken from a famous encyclopedia, summarized here. The specific domain of science studying life is called biology. You can study biology entirely, yet you will always remain confined to these four statements about life, acting as an official definition of life, while it is not a definition, and certainly not a mental model.

Clearly, these four statements attempting to define life do not form a genuine definition or mental model of life, but they only state some general information about life, simple characteristics observed in nature. Empirical definitions are observations of specific effects and characteristics, and therefore this definition of life offered by science is empirical, not cognitive, neither rational, nor comprehensive. If you want to know more about life, you certainly need an entire comprehensive mental model of life, since life is very complex. This is what we seek to accomplish throughout this book, a third level intelligent mental model of life, which is more complex than these four statements about life.

Why is this the case with the definition of life, one of the most important subjects to know? Is it because science is still young and ignorant, incapable to understand and explain life itself? Life is by far the most complex concept in the wider world. Furthermore, studying life is challenging, since you are bound to an only perspective of study, a unique point of view, found always within life, because you are permanently placed within life itself but never outside, and therefore you can never grasp it comprehensively from above. Because you can be nothing else but alive throughout life, and you cannot step outside life to study it wholly. Do not assume that you can exit life at death and become able to understand life entirely, from the outside, since it never happens. You are always alive, for as long as you exist in any form of life, since existence always defines you as a living being, while everything that exists in this world and in the wider world is always alive, in all possible types of life, forms of life, and classis of life, shifting at times from one to another.

Consequently, you are always bound to short loops of reasoning, keeping you in the empirical nature of your study, loops of reasoning blocking you from identifying and eliminating your beliefs and stereotypes about life, blocking you from achieving an accurate conceptual intelligent definition and understanding of life. This happens not only with life, but with all comprehensive supreme subjects of the world, as existence, the universe, intelligence, interconnectivity, cognition, causality, consciousness, conception, reality, and more. The current science can always study these at any level and in any manner, while it can always state consensually that it does or does not understand them, yet there is always a difference between consensual truth and accurate truth, and we must consider it accordingly.

Because if your need to know more about life persists even after reading the four statements about life given by science as a definition of life, then there is significantly more to know about life, while you must always know it in an intelligent accurate manner, otherwise you remain ignorant and undeveloped. Furthermore, since everything in life and in the world relates to life, if you do not understand life accurately and intelligently, in an entire accurate intelligent mental model, you cannot understand anything else accurately and intelligently in life and in the world, and you are doomed to remain ignorant, unintelligent, and therefore undeveloped.

There is a difference between accurate truth and consensual truth, since accuracy is unique, as it relates directly to the natural laws of the universe, while consensual truth is only an agreement formed by groups of people stating what should be true or false, good or bad, adequate or inadequate, allowed or forbidden, and moral or immoral. This is the case in all ideologies and jurisdictions including science with its entire scientific consensus, along with the entire mainstream society, since everything in society is consensual, but not necessarily accurate. The living, the dead, and the consensual.

Right now at the beginning of this study of life, you still have a choice, to study and learn everything either in an ideological manner, or in an intelligent accurate manner, since you are free to do as you please. You are free to accept consensual knowledge about life exactly as the scientific consensus and all ideologies offer, yet be ready to encounter a significant amount of contradictory information, because ideologies count in tens of thousands and they are always fighting for distinction and supremacy, while always doing everything possible to reach you. However, if you seek intelligent accurate knowledge instead, you must perform the entire study on your own, because the current science is also ideological, and cannot help you more. The current science is a scientific ideology, based on a scientific consensus, and cannot take you past its own set of scientific theories, which you already know well from school.

What should it be, the living or the consensual? It is tedious to reason at the intelligent human level, while you cannot use beliefs, stereotypes, and consensual knowledge throughout reasoning, since these are never compatible with intelligent reasoning and accurate knowledge. You must use only accurate knowledge while reasoning intelligently, since only accurate truth correlates and remains consistent in life and throughout the world. Accuracy is always interconnected, as it is integral part of life, of natural needs and meanings, and of the natural human environment.

Can you distinguish your own beliefs and stereotypes from among all accurate facts of your cognitive system? Which ones exactly are the cognitive structures that you can trust the most within your own reasoning? Because you must use your reasoning at its best in order to be able to understand life and the rest of the comprehensive subjects and concepts mentioned above. While many times, you cannot understand one without understanding the rest, since everything is connected in the wider world.

Let us now study this official 'definition' of life, one statement at a time.

1. Life is the characteristic distinguishing the living from the nonliving.

This first statement is the summary of a larger paragraph that I condense here, only to highlight its main idea. Here is the entire paragraph from your favorite encyclopedia:

Life is a characteristic distinguishing physical entities having biological processes (such as signaling and self-sustaining processes) from those that do not, either because such functions have ceased (death), or because they lack such functions and are classified as inanimate.

Do you see how, after discarding empty words and trivial remarks about life, you can have this paragraph in one line? Study it closely, to see how it is a cyclic affirmation and therefore it is a short loop of reasoning, claiming that living beings are alive while nonliving beings are not, with life simply distinguishing which is which, while making everything true in a trivial cyclical manner. Cyclical statements are always accurate because they are cyclical, and they can be used in any circumstance while always being true, as in straight lines always being straight, heavy objects always being heavy when they are heavy, or speed itself always distinguishing between fast objects and slow ones. Is this the only thing that you want to know about life, that the living is alive while the dead is not? You always want to know more, since this is why your curiosity persists after learning whatever the current consensual science offers about life.

Therefore, by giving this trivial definition of life, nobody knows what life is, nobody knows what this world is, while nobody knows how humans live, feel, reason, and why. While by not knowing these, nobody knows too much in this world, while everybody behaves as though they know everything in life and in this world, enhancing in this manner the dreadful human condition. Which means that, the current scientists counting in millions are just as smart, just as ignorant, and just as common as the rest of the world, with the difference that they form a scientific consensus that validates themselves consensually, stating that they are pertinent and right but not accurately, only consensually. In this manner, the current consensual science monopolizes the accurate intelligent human knowledge while contorting it in any consensual manner, increasing the dreadful human condition in life and in the world.

We notice how our model of life keeps hovering at the confluence between the living, the dead, and the consensual, while this might seem irrelevant, yet we are still searching for meaningful inner and outer perspectives allowing us to understand life. In this manner, we might even have these diverse perspectives, but only if we are capable to understand the living, the dead, and the consensual for what they truly are and for what they truly make you know, do, and be in life and in the world.

What exactly is consensual in the world? Everything is consensual around you, the entire current human society. Consensual means agreement. Study all agreements closely, since these take you out of life altogether, while even helping you understand yourself as a living human being, in an incredible contradictory circumstance, but still possible, since you are out of life while you are consensual, always understanding life better.

For example, let us now agree that the sky is yellow. We can always do so, since we can agree on everything that we desire, even that the sky is yellow. The sky is not yellow, but we only agree that it is, for any profitable reason, agreeing in this manner on lies. Notice the accurate truth, that the sky is blue, and the consensual truth, that the sky is yellow, because we had just agreed that the sky is yellow, while you are also part of this agreement, splitting the profit. Yet people do not agree on the color of the sky, but on minorities, decriminalized drugs, harmonized taxes and entire secessions, since these are very profitable.

Furthermore, if the sky was already yellow, we did not have to agree on it, because it was already the case as an accurate truth. This is the case with all agreements, since they must always form on everything that is not the case in life and in the world, while forming the entire consensual human reality, and therefore while forming lies.

This might never affect you if it concerns harmonized taxes, yet when people agree what specific knowledge should be accurate and what should be inaccurate, it interferes with your entire intelligent cognition, it controls it from within while affecting your social interaction and your human meanings, while also taking you away from life, since you do not follow truth, accuracy, and life anymore. If you have no knowledge and understanding about life, you never realize that you are out of life with everything that you do outside life, while placing you against life altogether.

This might be harmless or irrelevant if it is only your case, yet with everybody similarly consensual, everybody remains against life throughout entire dark ages since this is mostly the case, while always remaining against life and while always harming life. Furthermore, as long as humanity remains consensual, it always makes tyrants possible, while these keep humanity continuously in dark ages, always against life.

There are two types of civilization, golden and dark. You have intelligence, meaning, fulfillment, life, freedom, and golden human ages on one side, or doctrine, consensus, servitude, addictions, tyranny, and dark ages on the other. Make your choice, while accurate knowledge about life, along with intelligence, fulfillment, humanity, meaning, and golden human ages can always help you decide intelligently, but only if it is accurate, not consensual, because through the entire consensual, all tyrants of the world will always determine you to choose everything they desire, as dogma, bondage, drugs, servitude, and tyranny. You agree with these for profit, and alongside everybody choosing the same, you keep humanity in dark ages, always against humanity and against life. While if you have no accurate knowledge about humanity, human ages, and life altogether, you will always choose dark ages, while always harming life and the wider world. The sky is not yellow, but once you agree that it is, it becomes yellow, but only in your consensus, while taking you outside life and the real world, right into the Consensual Matrix with its everlasting dark ages.

This is how all agreements are made, for everything that is not already the case, because if it was already the case, you did not have to agree on it, since it was already the case. Let us now agree that the sky is blue, in order to be accurate and intelligent for a change. Yet we do not have to agree that the sky is blue, because the sky is always blue, it is always the case.

If this example is too simple, study closely all agreements, laws, beliefs, rules, statutes, and regulations made throughout all ideologies and jurisdictions, as these count in trillions throughout the world, to see how they are not the case in life and in the real world. Yet since people agree that it is the case, it becomes the case, but only consensually. This is how everything is taken outside life and outside the real world, consensually, with you included, and this is the difference between the living, the dead, and the consensual. They even make a court around you in court, in order to take you consensually outside life and outside the real world, while it is impossible to take you outside life and outside the real world, because you are always an intelligent living human being living your life continuously in life and in the real world.

It is enough to distinguish the real from the consensual, in order to be able to identify life and your own meaning and fulfillment in life. In general, whatever you do naturally, you do for Life, while whatever you do consensually or artificially, you do on behalf of all tyrants of the Consensual Matrix from here and from everywhere else. Furthermore, by obeying the Consensual Matrix, you go against Life, and this is how you harm yourself, your loved ones, and the entire world, cloaked many times in beliefs, laws, corrupted knowledge, lies, and stereotypes.

This is why all four statements from the officially-accepted 'definition' of life do not reveal much truth, because the current consensual science is part of the Consensual Matrix, and it never studies life, since the Consensual Matrix is dead, consensually dead. In this manner, the very important accurate intelligent knowledge about life always remains hidden, one dark age after another, to make all tyrants possible.

Furthermore, a rigorous understanding of life helps you understand yourself entirely, including your higher self. Know yourself entirely, and nothing can ever control and exploit you. Because you, as a multidimensional being, are very capable, standing above tyrants, and they cannot reach you to exploit you while you are fully developed.

The current science hides important scientific subjects behind silly scientific remarks, while forcing you to repeat them word for word during exams and throughout scientific reports, as it does with life. The current science attempts to define life empirically, through everything that you observe and perceive about life, and not in a comprehensive intelligent manner, as you might expect from the official scientific domain of society. This happens because the current science cannot exit life to study it comprehensively, which is a mandatory perspective in every important general study. Whatever the case is, it seems that science never even attempts to study life rigorously, restricting itself to the empirical study of what we call organic life, while assuming it to be the only life on Earth and everywhere else.

Understanding life rigorously is imperative for knowing everything about your entire existence in the wider world, while by not knowing what life is, you fail knowing who you really are, and therefore you are being used in every manner by anyone and anything that happens to be of a different type of life, with and without your knowledge. Let us now study the second official statement about life.

2. The smallest contiguous unit of life is called an organism.

This is not a scientific accurate truth, but a scientific consensual truth, decided through scientific consensus. The smallest elemental living being is not an organism, because there are entire forms of life below organisms, filled with living beings so small, that you cannot even detect them. The lowest detectable form of life is the ionic form of life, standing at the base of the entire organic form of life, while making possible organisms from below. Cells are not organisms, because cells are of the cellular form of life and they are called cells, not organisms. It is erroneous to refer to prokaryotic cells as single cellular organisms, because cells are cells and organisms are organisms. Similarly, it is erroneous to refer to bicycles as two-wheel cars, because cars are cars and bicycles are bicycles. Furthermore, living beings are always alive, never units.

The smallest living being is not an organism, since even cells and cellular components are alive, of the ionic form of life as it is the case with all ions forming cells, of the molecular form of life as it is the case with all enzymes and proteins forming cells, and as it is the case with all cells, prokaryotic and eukaryotic. The current scientists had decided to accept that the smallest living being is an organism of the organic form of life in a consensual manner, similar to deciding that the sky is yellow. Furthermore, the rest of the world has to agree similarly erroneously alongside them, because the current science monopolizes the entire scientific domain of the current society, confining you in an unreal nonliving consensual world, the Consensual Matrix. While in this manner, when you are in court, they do not even have to make a court around you to state that you are in court, because you are always in a consensual court, with everything that you know consensually about life, humanity, and the real world.

The smallest living being of the organic form of life is an organism, while the smallest living being of the electromagnetic type of life is of the ionic form of life, and it is any living ion found in any cell, because everything ever forming Life is always alive, including all ions and all molecules forming all cells.

Furthermore, the essence of life itself is at the base of life, in the smallest living beings of life, as these give life to all forms of life and to all classes of life above. There is no specialized cell of pure life throughout organisms, but all cells are similarly alive, while giving life to the entire organism through their own life. Similarly, within cells, there is no cellular component specialized in pure life, but all cellular components are similarly alive, while giving life to the entire cell from their own life.

Because all cells within organisms resemble more to entire cellular societies or to entire cellular civilizations, with all their components as genuine living beings, while in this manner, the entire electromagnetic type of life has the ionic form of life at its base making possible the molecular form of life above, which makes possible the cellular form of life above, which makes possible the organic form of life with humans included, which should make possible the social class of life which is the living humanity, only that the current society is consensually dead, while the actual living intelligent human society that should have been the entire living humanity or the entire living human class of life is neglected, restricted, censored, and therefore impossible, ending in this manner the electromagnetic type of life for human beings with humans themselves, at the organic form of life. This is the case throughout all dark ages, only to make tyrants possible, marking the most significant human failure, and nobody cares. Because once you live your life drugged, tyrannical, consensual, and dogmatic, you never care, while you make possible all dark ages, one after another, and the tyrants love you.

The same consensus of scientists decides currently whether Pluto is a planet or not this decade or the next, while you flunk your exam and it ruins your life if you state the wrong answer. While when you study the planet Pluto closely, you even find a larger mountain formation resembling the cartoon character Pluto, and therefore everything is consensually true, that Pluto is a planet, that the smallest unit of life is the organism, and that the sky is yellow, since it is the same in the unreal consensual world.

What happens if you consider cells similarly alive as the organisms that they form? All cells are alive, regardless if they live life independently or in entire organisms. This is always the case, despite of any definition of life that you ever consider, because any consensus determining any scientific knowledge remains insignificant compared to Life herself, while all components of Life are living beings and are alive, giving themselves life to the entire Life.

All seven trillion eukaryotic cells forming your body are considered by science nonliving, while they are about twenty times larger and significantly more developed than the prokaryotic cells that the current science considers alive. Prokaryotes are the cells considered by the scientific consensus alive because they do not form organisms since they live life independently, as all microbes do. Yet the prokaryotic cells are considered currently single cellular organisms, since this is why they are considered alive even as cells, which is as stating that bicycles are two-wheel cars, always referring to bicycles as cars because only cars are vehicles. Nobody can fool you regarding bicycles, while everybody tries to fool you in everything regarding science, always diverting you away from the accurate scientific truth.

Because once you learn everything accurately about life, you actually start fulfilling the meanings of life as an intelligent living human being, while in this manner, you keep humanity only in intelligent golden human ages, with all tyrants impossible in this world. Yet since this is not exactly what the tyrants want, because Kim must always have his dancers, Xi must always have his money, Donald must always have his North America, while Putin must always have his wars, you learn only contorted consensual knowledge, keeping you undeveloped.

Furthermore, eukaryotes, these seven trillion cells forming your body, are not the only living beings participating in your existence. You have prokaryotes in your digestive system, bacteria helping you digest and assimilate food, while these prokaryotes are more numerous than the eukaryotes forming your body. You live life in symbiosis with them, trillions of bacteria found mostly in your intestines, while by any scientific law, they are considered officially alive, with all the rights and privileges of a living being.

According to science, prokaryotic cells are far more

superior than your fetus, which is considered officially nonliving, while whatever science states has priority in court, overruling everything. Science makes the laws and dictates society, exactly according to the consensual agenda coming from all tyrants of this world, and from all tyrants of the worlds above through the Consensual Matrix. Once the human consensual agenda is dictated from outside humanity, this is not a human world anymore, rendering you incompatible as a living human being in your own world. This is the dreadful human condition, consensual, taking you away from life and from the real world, one dark age after another.

When all cells from your body are considered alive, both eukaryotes and prokaryotes, then the human fetus has to be considered alive, a developing living human being. While as all human beings, the human fetus has human rights, equal to everybody. In this manner, abortions become murder of all degrees, and must be considered accordingly in court.

How would this change the world? Drastically. How important is life exactly in the world, human life, along with family life? Let us see, because when you study life intelligently, you must study it in all relevant circumstances.

Women have the right to decide whether they can have babies or not, since it is their own body, and they can do as they please. Is this correct? Yes, it is a right that everybody has, men and women alike, to reproduce or not, if they want to do so or not. Because reproduction is a physiological human need, and as it is the case with all needs, human, consensual, or animal, you always have the right to fulfill them in any manner you want, or even to ignore them altogether, if this is what you want. Therefore, you always have a choice, and therefore you can always choose if you want or not to reproduce. Which means that women have their women rights assured even legally and morally.

This is what men and women do when they engage in all sexual acts, they do everything in order to have children, since everybody already knows where children come from and how they are conceived. You sleep in order to recover, you drink water in order to remain hydrated, you breathe in order to intake oxygen, and you have sex in order to have children, since all these are basic physiological human needs, while they are very important to fulfill, otherwise there are no more humans, and there is no more human life.

However, nobody forces you to fulfill your needs, because you only have the right to fulfill your human needs here in this world, but not the obligation to fulfill them. You have the right over your own body, meaning, and fulfillment, and therefore all women have the right to reproduce or not, while they have the right over their own body, with nobody allowed to interfere.

Furthermore, the current consensual science considers the human fetus nonliving, since it considers that the human fetus is not a human organism, while it considers only organisms alive. Yet since the current consensual science has priority in court, it states that the human fetus does not have human rights since it is not alive, while giving to women complete authority over them, to do as they please.

However, this is only the consensual human knowledge, because accurately, all cells are alive, including the eukaryotic cells forming organisms, and including the human fetus. The human fetus is alive starting with conception. The human fetus is always a living human organism, it has human rights, and cannot be harmed nor killed once it is conceived.

This is the accurate living versus the consensual, with the accurate living fulfilling life accurately while making the entire human life possible, while with the consensual killing life legally and illegally, part of the comprehensive genetic extermination taking place on Earth during all dark ages.

Systematic genetic eradication, because you do not see members of the Higher Brotherhood and of the Elite throughout abortion clinics, but only members of the Masses and of the Lower Brotherhood, considered disposable of the current consensual society, with the comprehensive genetic extermination taking them out considered normal culling, since the Elite does not even consider the lower social classes human. Therefore, every time you sustain, promote, and perform abortions, you enter in agreement to eradicate humanity genetically, while maintaining this consensus with all tyrants of the world who do not even consider the world below them human, but only animal or monstrous, as abominations. Yes, the invisible kingdom considers humans abominations and golem, while eradicating humanity genetically continuously, through wars, medicine, ignorance, drugs, lies, trickery, food additives, homosexual interaction, systematic poverty, abstention, sex change, and abortions. While as you notice, every time you study life in a tyrannical undeveloped consensual world, you end up studying death persistently, one dark age after another.

Are abortions good or bad? Are human abortions the human freedom itself? By having the right to reproduce or not, you have the right to engage in sexual activities or not, in order to fulfill your living need for reproduction. Once the conception is successful, that first human cell, along with the entire human fetus throughout gestation, that fetus itself is a developing living human being, which is your child, part of your family and loved ones, since all children are developing human beings. You never have the right to kill a human fetus, because it is not you, it is not your body, it is not part of your body. You have rights only over yourself, over your own body. You always have the right to reproduce or not, by engaging or not in sexual activities, but once you do, and once your child is conceived, you can never kill him or her, since it is not you but someone else, a living human being, and if you kill him or her, it is considered murder.

Should you also have the freedom to kill your other children at home, along with your siblings and parents? No, since the current consensual society does not allow it, because it is considered murder, and it is dealt with accordingly. Additionally, Life herself does not allow you to kill others, exactly as you feel it through your human needs, since this is why you never kill your loved ones, but you always love them. The current consensual laws of Earth are implemented and maintained in order to protect humans and human rights under all circumstances, only that all laws of Earth are contorted continuously, in order to be used for profit, through discrimination, exploitation, and extermination, as it is the case with abortions themselves, because this is the difference between the living and the consensual.

Which means that there should never be debates to legalize or not abortions, because the human fetus is a living human being having all human rights, and any harm done to it must be considered legally accordingly, as it is the case with all human beings. All abortions are human murder, while human murder is undebatable, always punished in court, making all abortions undebatable, always illegal and always punished, since human murder is always illegal and always punished.

This is called reasoning accurately at the intelligent human level, yet you never find this type of intelligent reasoning in the world, but only debates, along with a large amount of consensual knowledge, but not accurate intelligent knowledge and accurate intelligent reasoning. The entire world debates if abortions should be legal or not, for various reasons, circumstances, and consequences, while never considering that abortions are already murder, while murder is already illegal in the current society. Only that, in the current human society, it is considered consensually in a contorted manner that the human fetus is nonliving, until birth. The sky is yellow. This is how the consensual ends up deciding between the dead and the living, which should never be the case in an intelligent human world.

Is the human fetus actually alive? Yes, since it can always live its life independently from the mother, in incubators if necessary. Which means that mothers do not give life to their children after conception, because they are already alive, and already human. Fetuses are considered nonhuman and nonliving only consensually, either legally or ideologically, ending up dead most of the time, aborted. More precisely, people decide through agreements that humans are nonliving and nonhuman throughout gestation, which is not accurately true, just the way people can decide that the sky is yellow through similar consensual agreements, which is not accurately true, and this is the difference between the living, the dead, and the consensual. With the distinction that during golden human ages you have only the living, while during dark ages you have only the dead and the consensual, because every time you are not part of the meanings of Life you are dead, consensual, and nonexistent.

If this is still inconclusive to you, the consensual always remains compatible with itself on its own, in a major conflict of interests, forming by law, beliefs, and agreement an entire Consensual Matrix covering most of the wider world, while exploiting systematically all life of all types and all forms throughout all realities of the wider world, intelligent or not, as it happens on Earth during the dark ages.

Currently on Earth, many courts and jurisdictions consider the human fetus alive but only after a specific number of weeks, while in this manner, Life herself is decided consensually to exist or not in the world. Life herself, who is always here, decided consensually, in the world, by living human beings, as being or not real, as being or not true, as being or not the case, and as being or not alive. Life herself, and this is what you are always dragged into, consensually, while worsening the dreadful human condition.

According to which laws do women gain human rights against their own children? What exactly are the human rights, and who defines these? All human rights are recognized through higher laws. This is the case because all laws promoted and enforced in society throughout all nations of Earth apply only within jurisdictions, but never in the real world, and never on living human beings. Because all jurisdictions, ideologies, and hierarchies are consensual, fiat, not there, and therefore they can never address living human being, while this is what they also claim themselves, through laws and agreements, in a major conflict of interests. This is why you have to be a consensual corporation in order to fit in any jurisdiction, in any court, and in any system of justice, or this is what they make up about you, consensually.

Furthermore, ideologies, hierarchies, jurisdictions, systems of laws, systems of justice, and entire consensual districts, states, and countries cannot define and assign human rights to humans, because humans are always alive and they always have their human rights. While these social organisms are consensual, always under-standing humans, and cannot decide anything above their own legal and ideological jurisdictions.

This is the difference between the living, the dead, and the consensual, with both the dead and the consensual being nonliving, or this is what it is agreed upon consensually. Laws, beliefs, statutes, and privileges apply to all consensual corporations, all trademarks and brands, while these cannot have children, since they are consensual but not alive, while taking away from life and from the world the real living human children.

Who does all these terrible things? The demons, the devil, and the aliens never harm you, or not directly, because you do, yourself, since you kill your own children yourself. You do so in order to improve your actual human condition in life and in the world, since this is what everyone does, or this is the stereotype. Because once they control the stereotype, which is consensual, they do not need demons and aliens anymore to harm the human world, since they have you to harm yourself and everybody else. All higher laws are implemented exactly to protect all life in the wider world, including intelligent life, yet with you harming yourself, your loved ones, and your entire world, all higher laws stand by, while respecting your own rights, status, and decisions in life and in the world, even your choice to harm yourself, your children, those around, all life, and the entire world.

You already know the higher laws, not only from justice, education, and society in general, but from religions and schools of thought, since they are given many names, as divine laws, divine teachings, karma, and commandments. The higher laws are not similar to the natural laws of the universe, since higher laws are implemented through higher consensus, with the purpose of creating order throughout the universe, at all living levels.

The natural laws of the universe define this world along with many other correspondent realities above. The natural laws of this universe derive furthermore into the laws of mathematics and physics, being structural part of the Field, Existence, Life, and the universe, applied naturally, regardless of any wish or consensus ever formed. The blue color of the sky is formed by the visible light coming from the Sun when it enters the Earth atmosphere, making the sky blue through the natural laws standing at the base of this world, in an accurate manner. Not only the color of the sky, but everything accurate is determined directly through the natural laws of this universe, because the natural laws of this universe define directly all events, concepts, circumstances, and all lines and lifelines of causality of this world. While all natural laws of this universe had been decided by our Creator when he created this world, defining accuracy altogether in this world ever since.

You do not need consensual authorities of truth deciding the truth from the erroneous on behalf of all tyrants, because the natural laws of this universe always define and determine everything accurately true. You cannot change accuracy from within this world through any consensus that you can ever adopt, yet when humanity entirely unites against our Creator through all consensual jurisdictions and ideologies, it has a decisive impact, with you caught right in the middle.

The natural laws of the universe are accurate continuously throughout this entire world, otherwise this entire world is not possible, since they stand at its base. Therefore, no one can break, evade, or contort the natural laws of this universe, just the way you cannot break the laws of mathematics, classical physics, and chemistry, since these relate directly with the natural laws of this universe. Unless you know how to bend them, through higher powers, but even then, you can bend them only here in this world, and only locally and temporarily, depending on various circumstances.

You can base all your acquired knowledge on all natural

laws and higher laws, here in this world and in all worlds and realities of your study, since Life is everywhere. While all you have to do is to distinguish between the living and the consensual, between the real and the consensual, or between the natural and the consensual, since it is the same. Because once you consider the living, the real, and the natural, you have these natural laws at the base of all realities that you study, as your living, cognitive, real, or natural realities, helping you remain accurate and consistent throughout the study. While you can achieve consistency and accuracy throughout all studies, if you are capable to detect, remove, and avoid the consensual itself, which is everywhere during the dark ages.

Right now, you might have a multitude of beliefs, strong personal convictions, and stereotypes contorting your mind, as it is the case with everybody else, since the current society is consensual, as the Consensual Matrix attempts to make all intelligent life consensual throughout the wider world. Be well prepared in a consensual undeveloped world, because in general, people use less than one percent accurate cognition throughout life, with the rest being consensual, erroneous, unreal, inadequate, intoxicated, unstable, indoctrinated, sick, or malfunctioning, always affecting you.

The Consensual Matrix does not consider humans alive before it is capable to register them as brands and corporations, in order to have them for itself entirely. Once children are born, parents demand a Certificate of Birth, and this is how children become part of the Consensual Matrix, by law. Check now your certificates of birth, and if they have the names in uppercase letters, those are corporations. These certificates never address living human beings, but only consensual corporations, registered and therefore instated exactly at birth, with the date of birth. Which means that, as a living human being, you can be in this world through human rights and human status, and never by consensual agreement to be here, as permits and certificates. If you do not even have your certificate of birth, but only copies, it means that others have the originals since they own your corporation, as this remains the case with entire nations many times.

Living human beings cannot own and cannot exploit living human beings in a democratic world, yet all laws can always be contorted in any world, even throughout democratic worlds. Living human beings cannot own or exploit living human beings, however, corporations can always own and exploit other corporations, as it is always the case at work and in all other jurisdictions. If you do not have your certificate of birth but only a copy, it is more likely that other corporations own your corporation by default since birth, exploiting it in any consensual manner. Even if you have your certificate of birth in original, other corporations can always own your corporation through employment, loans, debts, oaths of servitude to masters and royalty, military oaths, public service oaths, and through all possible ideologies and jurisdictions reaching you. Because if you are in an ideology or jurisdiction, they can always consider consensually that they own you.

Let us now study the third statement from the official definition of life.

3. Organisms are composed of one or more cells.

This statement reinforces the previous one, that only organisms are alive, not cells. Is science so ignorant to state the same thing twice, or there is more to consider? Let us put the second and third statements together: *'The smallest contiguous unit of life is called an organism, while organisms are composed of one or more cells.'*

These statements, together, strengthen the consensual agreement that only organisms are alive, with everything else lacking the status of a living being. Is this important? Yes, since not only that this scientific law allows abortions, deciding consensually which genetic lines to live and which to die, but this is how the official definition of life considers only organic life to be alive, and nothing else. In this manner, your questions about spirits and life after death seem very silly, while they do not apply to science by definition, but only to fiction, religion, paranormal, and philosophy. Because science does not offer the necessary scientific answers throughout the development of our civilization as it should. Through its false, preordered and concealing scientific laws, science interferes and alters rigidly and deliberately the development of our civilization, in a harmful manner. Because a civilization lacking overall awareness of its own environment has a lower status, and therefore it can be used and exploited by beings, entities, collectives, and entire civilizations throughout the Consensual Matrix. Civilizations are considered alive at a higher level of organization, just as all societies, ecosystems, and biospheres, since they are part of Life. However, entire civilizations can be incorporated consensually with a simple statement, agreement, signature, and registration number, and they become part of the Consensual Matrix.

Let us continue with the fourth and last statement of the official definition of life.

4. Organisms undergo metabolism, maintain homeostasis, grow, respond to stimuli, reproduce, and adapt to their environment in successive generations.

This is certainly true about organisms, and we knew it well since primary school. However, some of these simple remarks do not apply to organisms as it is clearly stated, but to all types of life, forms of life, and classes of life. While the statement of organisms undergoing metabolism is too vague even to differentiate the living from the nonliving. The statement of organisms maintaining homeostasis might be specific only to the electromagnetic type of life, and it might be the same with growth. Reproducing forms of life naturally grow, since they are not immortal. These are characteristics of organic life. Organisms and life in general respond to stimuli and adapt to their environment. Yes, this statement should apply to all life, and we keep it for our study, yet we already knew it since primary school.

This is the 'definition' of life given by science. Is this enough to get us started in our own quest of learning everything about life? Let us see. The first line of the definition of life is trivial, the following two lines are consensual knowledge and they apply only to organic life, while from the fourth line, we can use the well-known observation that life responds to stimuli and adapts to the environment, which is far from sufficient to describe life in its entire living complexity.

Let us now state the human limitations while understanding and defining life, because we attempt to define a tenth level supreme living being, Life herself, through first level algorithmic statements and definitions, since only first level algorithmic concepts can be communicated throughout books in words and paragraphs, as it is the case with all written communication. Furthermore, the human life is of the third intelligent level, matching the third level intelligent human reasoning allowing us to understand it, while also matching this entire third level intelligent mental model that you form in your mind right now as you read this book, allowing you to understand life as much as possible. In contrast, everything consensual is of the first level, easier to define and understand, while all animals are at their second instinctual level, also easy to understand through your third level intelligent conceptual reasoning.

Which means that you can always understand life, Life, nature, reality, and the wider world, but only at the maximum third intelligent level, through third level intelligent reasoning and mental modelling, if you are successful conducting these. Therefore, expect more from all higher concepts exceeding the third intelligent level, including life, wider world, higher cognition, higher meaning, and higher causality.

The third level intelligent model of life is not here in these words, but only in your mind, through your continuous intelligent reasoning that you perform alongside this book, if you are successful to follow this book. Yet if you have managed so far in the book without being bored or left behind, then you reason intelligently alongside this book at the third intelligent human level.

From the start, we notice separate concepts embedded in the word 'life.' The most common concept is the one defined by science above, the concept of life differentiating the living from the nonliving. The second concept of life means life, as the life in you, the essence or component animating you, making you alive. A third concept of life relates to vitality and energy, as in how much life you have left in you, or how lively you are currently. A fourth concept refers to life in general, organic or not, developed or not, terrestrial or not, and this is Life, while she is alive herself, constituting all life in the wider world.

Life is alive, you are part of her, and you contribute to her through all your intelligent human needs and meanings that you fulfill successfully throughout the day and throughout life. This is in contrast with the multitude of consensual needs and meanings that you fulfill for the Consensual Matrix. The Consensual Matrix operates many times against Life, tricking you and constraining you in every manner to undergo an entire harmful consensual operation against life and against everybody else. The current consensual society even encourages you to seek profit in a continuous social competition against life and against everybody else, and since you remain continuously against life, it is a comprehensive funeral march taking humanity to its common grave, with everybody marching carelessly and undisturbed. This is the dreadful human condition, the actual human burden that you carry with you throughout life knowingly and unknowingly, manifesting throughout all dark ages, only to make some tyrants possible.

As stated, there is not only one concept of life but four, because this is how life is considered conceptually, ending up with four homonyms of life. You can certainly force your understanding of life to adapt these four concepts written as life into a single definition of life, even if you have to derive them one from another as science does, yet it would be erroneous, triggering scientific stereotypes, because each concept of life must have its own accurate definition, in agreement with the natural and higher laws. These are very important distinct concepts of life, they are neither complementary nor synonymous but they are homonymous, and they should be studied and understood separately. These concepts are so distinct, that they should have their own words, and not the same word, life.

It is very important to understand what differentiates these four concepts of life using the same word life, and why it is so important to understand and assimilate them separately, even if so far, with our common understanding of life, they seem trivial and almost similar.

If these four concepts are so important and so distinct, why are they used with the same word, life, and not with four separate words? Because vocabularies are formed naturally, by the general population, and not exactly consensually, but naturally, through living human interconnectivity, matching perfectly the understanding that people had about that world at that time, while matching the need for the specific words that they had at that specific time. Furthermore, vocabularies are transferred from one generation to another for decades, centuries, and millennia, while they are very rigid to change, regardless of the change in the understanding that the general population has about life and about the world. This is how, many times, several words remain associated to the same concept purposelessly, while other times, several concepts remain associated to the same word. Vocabularies are so rigid to change, that even those who control science and society have difficulties in changing words, concepts, and meanings. This is why etymologic studies reveal more truth than expected, and they are very helpful.

For example, the Inuit people have fifteen different names for snow, while these are synonymous from our own perspective, or not exactly. One means squishy snow, another mushy snow, another squeaky snow, and another icy snow. It is mandatory that the Inuit use the right word for snow in their conversation, because each word defines a different concept of snow, which is directly related to the weather conditions current and predicted. While if the weather conditions do not matter too much where you live, it certainly matters in cold regions. Squeaky snow is sharp, not too slippery, and it means minus twenty degrees Celsius and no snow for the next four days.

The following two examples are from religion. People of some religious background use four different words to define the same religious character: Satan, Devil, Demon, and Lucifer. It has always been in this manner, and it is appropriate to use these words as synonyms. Is this true, or there is something else behind these specific words?

If your religious background is too strong and you do not find the use of these words as synonyms too remarkable, then let us take another religious example, from the Ancient Egypt. Several thousand years ago, people used three names together to define their one deity: Osiris-Ammon-Ra, for only one deity. This is unusual for us, since we know that these are three distinct major religious characters, they cannot refer to the same deity, and therefore they cannot be synonymous, because they are different deities. Yet if you lived there and then, that was your deity, those were your synonyms, and you never thought much of the names. This might be similar with you currently while using several names to refer to the devil, never suspecting that there might be more than one devil in the world. Can it be more than one?

Why does this happen? Because those in control, whenever they change everything consensually on their behalf, have difficulties changing the vocabulary. Do a research, to see how all religious characters are either good or bad, depending not necessarily on their actual behavior, but depending on the time, place, nation, culture, authorities, ideologies, and regimes, with people themselves deciding consensually who is good and who is bad, regardless if these characters were good or bad themselves. Seth was a major Egyptian deity before Osiris and Horus, while Satan means follower of Seth. 'Devil' means 'he who protests,' the way people protest currently in the street during riots while frustrating authorities or while changing them altogether. 'Demon' is the demi-Ammon, the direct descendant of the deity Ammon on Earth, a great ancient deity, word related currently directly to 'amen,' which is commonly used at the end of prayers in many religions.

'Lucifer' means 'the light bringer,' or the teacher of higher truth and higher knowledge, which he had brought at high risks from the higher worlds. Are you a seeker of higher truth and higher knowledge? Do not rush up to judge yourself through all negativity associated consensually with Lucifer currently, because there was a time when Lucifer was the hero, the pride of the ancient times, when it was said that Lucifer had risked everything to descend to Earth in order to bring higher knowledge to humans, and therefore his name Lucifer. While as you study closely all religious characters, you find them positive and benefic during one age, and then they are considered the opposite during the next age, very dreadful. This is the difference between the accurate and the consensual. Accuracy is unique, determined directly by our natural laws of the universe, and it cannot be otherwise here in this world. While the consensual can be everything that people decide throughout ideologies and jurisdictions in an inaccurate manner.

You do not have to acclaim or disclaim these religious characters, since you do not have to invest your effort interpreting them as they are depicted currently consensually. Religious truths and religious beliefs are consensual, and therefore they are changed repeatedly throughout time through religious consensus, just as science and education change everything according to the same higher consensual agendas.

Who changed these concepts so drastically? The people controlling and exploiting you, while life on Earth has always been in this manner throughout the dark ages. You were there then, you are here now, as you are certainly coming back for more, since it matters less where you live and what you do if you are in the Consensual Matrix, as long as you feed the Consensual Matrix.

Why having four different concepts for only one word, life? You are familiar with the anecdote with the three blind men trying to define an elephant. One touched a leg and stated that the elephant is a tree, another touched the trump and stated that the elephant is a snake, while the third man touched the tail and stated that the elephant is a rope. We are also in this circumstance, because our limited senses of perception stop us from seeing and understanding comprehensively the true life at her tenth supreme level, keeping it hidden from us, while our confinement to this world stops us from seeing and understanding spiritual life accurately. Furthermore, the fact that we live in a single place and on a single planet stops us from seeing and understanding other types of life, while the fact that we measure our life in seconds, days, and years stops us from understanding very fast and very slow types of life. Our only scale constrains us to perceive and understand only macro life, while the fact that we are always alive stops us from forming a general understanding of life, enough to be able to make the general distinction between the living and the nonliving. Furthermore, science does not help much, and therefore you are on your own throughout this entire very demanding research.

These constraints are certainly unfortunate, yet our circumstance is not too bad, since once we understand, realize, accept, and reveal our limitations, we can always find a way to answer accurately all our questions. These four concepts of life are the result of the inevitable dissociation that we encounter while trying to observe and understand life from our unique perspective within our unique circumstance, just as in the anecdote with the elephant and the three blind men.

What exactly differentiates the living from the nonliving? Can we call life simply 'animation,' as science does? Here is a cat on the table, alive, jumping everywhere while playing nicely, and here is a cold heavy rock on the floor, always motionless. Why is one alive and the other nonliving? Because it is certainly not only animation differentiating them, since my current bicycle cries as a cat sometimes, and takes me everywhere in a very animated manner, yet it is not alive. It seems very easy to understand life, yet it is not.

Furthermore, you cannot differentiate entirely the living from the nonliving. The cat is not entirely alive, since its bones might contain some living cells, yet they are mostly made of a solid structure of calcium, oxygen, carbon, nitrogen, hydrogen, and silicon, and it is the same with its claws, hair, whiskers, and cornea, since these seem less alive. Even its cells have nonorganic components in them, and when you study them closely, all elementary particles, ions, salts, and small radicals seem less organic and therefore less alive. All cellular components are of the cellular form of life, molecular form of life, and ionic form of life, but where exactly do you draw the line between the living and the nonliving?

Where is life? It seems that the living merges with the nonliving, while everything is alive, including the cat and the rock, only of different types of life and forms of life. However, molten iron at the forgery might not be too alive, while you can find life even in rocks, stones, and mountains.

If you want to understand life and the living, you must understand the nonliving, while the nonliving is very hard to find, since it is only in molten lava and in molten metals, wherever the electromagnetic type of life cannot reach. Yet even there you can find life of other types of life, including in stars, in deep ice, in empty space, and in fire.

What are the living and the nonliving? Everything is alive in the wider world, to the point where Life is the wider world altogether, along with Intelligence and Interconnectivity. Even the meat at the grocery store is alive, not of the organic form of life anymore, but of the cellular form of life, otherwise you do not buy it. Yet even if it smells bad after several weeks, that meat is still of the cellular form of life, in form of microbes, since this is why it smells bad. That meat will live life at the cellular level with countless of living beings feeding on the original meat and then on themselves for many months until it dries up entirely, to become of the molecular form of life, which is still alive. Until is eaten by cells and organisms eventually, to reenter the cellular form of life and the organic form of life, for yet another cycle, which is very common in nature.

As you study the grocery store closely, which is the actual life from the human food chain, you find all food still alive if it is not already cooked, processed, semi-processed, or pre-processed, and therefore, you cannot define life through its empiric characteristic of thinking and moving around. All organisms who die shift to the cellular form of life and then to the molecular form of life and or ionic form of life and vice versa, depending on all possible circumstances, while as you notice, the matter itself does not die, but it only shifts from one form of life to another.

Can you still define death and the nonliving anymore? Yes, since death is only the downgrading of life from one form of life to a lower form of life. Yet can you still define life in an entire living world? What is life exactly, if everything possible is always alive, and you have nothing to compare it with? Is life still the characteristic distinguishing the living from the nonliving if everything is alive? You can define the organic form of life because you can always distinguish it from the cellular, molecular, and ionic forms of life, while you can always define the entire electromagnetic type of life if you can understand how it is made possible by the electromagnetic field, yet you might not be able to define all life, of all types of life and of all forms of life. This is important, because right now we try to define all life through more than the specific characteristic of life of moving around. We want the essence of life itself, since it defines life more than anything else.

Let us study the rock closely. The rock is full of life, and thriving. There are not only billions of microorganisms all over the rock, thriving, but there are more living microorganisms inside the rock, inhabiting every fracture and cavity of the rock, as small as these can be. Additionally, lithotrops are everywhere inside the rock, which are bacteria of the cellular form of life, feeding directly on rock. There is also nanolife everywhere within the crystalline structure of the rock, at the molecular form of life, making our apparently lifeless rock just as alive as the cat.

Let us widen our perspective, to make it comprehensive. Let us give our blind men arms and hands as large as the elephant, just to see what they can perceive from comprehensive perspectives. They certainly wonder where the trunk and the snake went, since there are no distinct trump, legs, and tail anymore, but only the entire elephant. Not only that their objects of study had disappeared, as the trunk, the snake, and the rope, but something else appeared in their place, the elephant itself. Because now, with arms and hands as large as the elephant, they are able to touch and perceive everything at once. The blind men will remain puzzled endlessly of where the distinct bodily parts went, while from our own outside perspective, it is easy to observe that the elephant is not a tree trunk, neither a rope, nor a snake, but a larger, comprehensive living organism, including all these parts and more. This is how all elephants should be perceived and understood, wholly, not one leg at a time.

It seems that whenever we study something important very closely, we tend to drift further away from its complete comprehensive understanding. Does this mean that we are merging further away from the answer? No, since it is a strategy to identify all beliefs, stereotypes, errors of reasoning, and limitations from our conscious and subconscious reasoning, eliminating them in this manner right from the beginning of our research, so they never alter our results. Because all consensual beliefs, stereotypes, and scientific consensus are of the first consensual level, all logic is of the first algorithmic level, all intuition is of the second animal level, while only reasoning, with its intelligent, rational, and conceptual thinking are of the third intelligent level. We can certainly use first level logic, second level intuition, and third level intelligent conceptual reasoning while mental modelling this entire topic of life, but we have to avoid the consensual, otherwise we end up with the well-known ideologies and consensual science that we already know.

Stereotypes are large chunks of information that we assimilate wholly throughout life, unconsciously, because we are in a hurry, because it is easier, or because everyone around does so or understands things in this manner, which is what science also does. Each stereotype, once identified, should be broken down into distinct elementary concepts, and then assimilated back into your conscious mind through elaboration, in an elucidated, logical, intelligent, conceptual manner. We have already done so with the lies and wordiness of the current science and with the four concepts of life, while we would have never advanced our understanding of life if we had not identified them first. The hidden stereotype relating the nonliving with the dead that you might still have is just another example, and I will point it out in a future paragraphs through a simple correlation to Frankenstein and his sewing machine. Furthermore, the stereotype relating the essence of life with animation will show up in the next chapter.

You certainly cannot understand life in general and your higher self in particular, if you do not identify all stereotypes loaded in your conscious and subconscious mind by science at school, by society in general, by media and entertainment throughout all movies, music, and news, and by religion throughout history.

We can certainly distinguish living beings from objects, since this is not a problem, while we always do so subconsciously, stereotypically. The problem is identifying how we do everything. Are living beings alive because they move around? Not necessarily, since plants stay still and are always alive, at the organic form of life. Are animals alive because they feel warm to the touch? No.

Let us now identify exactly why living beings are alive, and what is the specific living essence making them alive, if there is such an ingredient or essence to life making living beings alive. We experience this difficulty of distinguishing the living from the nonliving because we are incapable not only to understand the ingredient or essence able to render everything alive, but we are incapable to perceive, distinguish, identify, and isolate this essence altogether.

What stops us? We encounter more than stereotypes this time. You have learned to distinguish the living from the nonliving one object and subject at a time throughout childhood, while learning how to care for the living and how not to get hurt by nonliving objects while handling them around, forming a powerful stereotype in this manner. Simultaneously, throughout your studies and then throughout your entire life, science and society manage to enhance this stereotype with erroneous knowledge about life and death fed to you more or less deliberately, along with how you should feel and react in all circumstances.

Major stereotypes are very hard to identify, while if you ever want to identify them, you must do everything from scratch, even by reinventing the wheel, yet wheels are easier to reinvent than the entire intelligent accurate mental model of life. We still want to define life, through something more consistent than the current statement that life distinguishes the living from the nonliving, while if we also find the essence of life it is even better, since it distinguishes life more than anything else.

Let us now build or create a natural living being from scratch, since this will certainly help us remove all stereotypes about life, while helping us identify the essence of life itself. Because while building a natural living being, eventually, we will have to add this essence of life to our work of creation, this essential living ingredient that we cannot pinpoint at this time, which will render our construction alive. Let us take our time now to build this model of a living thing, and therefore let us see what exactly differentiates the living from the nonliving.

Frankenstein had managed to do so before us, while attempting to create life similarly, yet he had associated death with the nonliving, which is another erroneous stereotype, mostly when you consider only the organic form of life alive. Therefore, Frankenstein used the cellular form of life in order to create the organic form of life, by sewing tissue together, yet he did not create life, but he only uplifted life from the cellular level to the organic level, in a fictional manner, with dreadful fictional circumstances. We will not use tissue and sewing machines, but we will use robots that we can build ourselves in our garage, while always seeking to find a magical living component rendering them alive, the essence of life itself.

Life

The dead is certainly categorized as nonliving by science, yet it might be wrong, because the current science considers only the organic form of life alive, not the cellular form of life, molecular form of life, ionic form of life, nor the social class of life. The consensual itself is closer to the nonliving than expected, yet the consensual is also nonexistent, fiat, not there, never the case.

Death itself is not the nonliving, but it is only a characteristic. When you study death closely, you find it as an invention or ability of Life, not as the end of life or the other side of life, as everybody expects. Life had to invent and use death because she wants all her living beings to be meaningful and successful, while she wants to remove continuously the inapt, the old, the meaningless, the unfulfilled, and the unsuccessful from among all her living beings, through death. Which works fine, helping Life develop continuously throughout all her worlds and realities, even from one form of life to another either harmoniously or through the musical chairs scenario.

Do not underestimate Life, since she is highly complex, while death is not the only ability and characteristic of life. Life is highly intelligent and very harmonious, treating all her living beings and intelligences in a unique, priceless, fulfilling manner, while rewarding them greatly when they are successful, yet always killing everybody eventually, which is a major characteristic of life. Yet at least you are priceless and unique for as long as you last.

You are priceless, unique, and irreplaceable in life, tending to Life through your meaning and fulfillment in a unique manner. Because you are always who you are in Life, you are unique and alive, and you are not only what you do in life, as it is the case with all robots, servants, and machines. This is the difference between the living and the consensual, since you are always meaningful, fulfilling, priceless, unique, and irreplaceable in life, while you are disposable, meaningless, and unfulfilled in the Consensual Matrix, always having to take drugs to compensate, and always feeling worse.

Let us now create a living being from scratch, only to be able to see how life is and how it feels, while always searching for this major characteristic or essence of life, if we can ever find it.

2 THE ESSENCE OF LIFE

It seems at least amusing to attempt to build a model of a natural living being right now, when we are still in the beginning of this model of life. However, we truly have to understand the essence of life, what distinguishes the living from the nonliving, otherwise we risk doing this entire tedious research while ending up learning only about life, as biology does, but not exactly understanding life itself, as we intend. What can it be more meaningful to know than the essence of life itself, along with the origins of life, development of life, forms of life, structure of life, creation of life, purpose of life, and ending of life? Here we start, and we have only the space of this chapter to accomplish it, to find the essence of life along with anything else, so good luck to us.

Let us consider for now that animation itself distinguishes the living from the nonliving, as everyone assumes stereotypically anyway, in order to maintain consistency. Let us consider that life is mostly its own adaptation to the environment, as science assumes empirically. Let us now employ any available tool, device, scientific law, theorem, and concept, everything that our civilization can offer, to build our living being from scratch, because if we succeed, we find the essence of life.

If animation is all that it takes, then let us build a little robot, and let it run and jump around in a little circle. The robot moves around on its own, it is animated, and therefore it is alive. Yet not exactly, because we have successfully built an automatic mechanical device, which is certainly not alive. Our world is full of these machines, clocks, tools, and devices, while they are only objects, never alive. We failed in only one paragraph, because science and everybody else failed understanding life, while we only maintained consistency.

Automatic devices and machines move and behave in a predefined repetitive manner, while they require energy and constant maintenance to do so, yet I still refer to this as first level artificial life, because machines and computer programs still exist and function well even on their own, while they are not alive. This is artificial life, either mechanical, electrical, or digital, while genuine life starts at the second level, with cells, cellular components, and entire plants and animals. Examples of first level consensual, digital, and mechanical artificial life are corporations, machines, clocks, computer software, and computer applications, while they have a first level artificial intelligence, which lacks its own meaning and consciousness, all being used as tools of all kind, in a repetitive, algorithmic, or consensual manner.

We persevere, as we study organisms. Living organisms take energy from food or from the Sun, and they constantly maintain themselves. Therefore, this is what we must have in order to build life, a robot equipped with solar panels, able to maintain itself, and now this is how we create life.

We have already created life at a very low level, at the first artificial level, while this is artificial life, or this is how it is called. This is not natural life, since it lacks exactly what we search, the essence of life, and it cannot help us much.

Yet it still helps us, since we now know that the essence of life is natural, it relates both to the physical body and to intelligence, these have to be natural, and more importantly, these must have a normal natural consciousness, choice, autonomy, and survival. This is what we want to create, a

Life

natural living being, not an artificial one, and we want to do so from scratch, because Frankenstein used sewing machines, and had all those problems.

Levels of life are different than types of life, characteristics of life, laws of life, forms of life, perspectives of life, or existential types of life. Levels of life define life through a generalized interconnectivity in life, which can take place only on ten distinct resonant harmonious levels, from zero to ten, with all living beings of all life interconnecting only at these specific resonant levels. Animals are at the second intuitive level, humans are at the third intelligent level, nonexistence itself is at the zero level, the consensus is at the first consensual level, while Life herself comprehensively is at the tenth supreme level. Our robot is still at the first artificial level, it is not alive because we had not found the essence of life yet, and we must persevere.

You cannot state a simple characteristic of life in order to define life. You must always consider more in order to define life, not only its characteristics, but also its types, existential levels, laws of life, forms of life, and perspectives of life. Therefore, we must consider all these while building from scratch a living being, while as seen, we had already built a first level algorithmic artificial being, by building a normal robot. Similarly, we can always build zero level life, by making any object, since these are of the zero level life. Yet since our robot moves around, we consider it alive at the first algorithmic mechanical level, similar to all clocks, machines, and automated toys.

The consensual is at the first level of life, similar to the algorithmic, yet this is not the case in real life, but only consensually, by agreement, similar to the yellow sky. Otherwise, the consensual is only at the zero nonexistent level of life. All corporations that the consensual instates by the trillions are neither real nor alive, but they are only considered consensually real and even consensually alive, yet in a fiat manner, not exactly there. If we want to create a consensual living being or corporation, we can simply register one legally since it is very easy, yet this does not contain the essence of life.

We are already successful creating artificial algorithmic life of the first level, by controlling objects directly or through simple algorithmic procedures as *if – then,* and *repeat – until,* yet this does not contain the essence of life, since the first level algorithmic artificial life is only robotic life, computer life, or automated mechanical life, while normal natural life starts with the second intuitive animal level. The machines from all factories are of the first algorithmic mechanical artificial level of life, since they repeat the same algorithmic movement throughout all assembly lines endlessly. It is the same with the workers working on assembly lines, as long as they involve no reasoning throughout their work, but only repetitive procedures while aiding the machines. These are first level behavior, first level thinking, and therefore first level life, artificial, algorithmic, and or consensual.

It is true that the people working on assembly lines are of the third intelligent level since they are living human beings, yet only their behavior and interaction is of the first algorithmic level, which becomes their level of life for the human society and for life altogether, very low. This was the case only in the past, because currently, all the workers of the assembly lines must involve their intelligence, since if they involved only their basic first level algorithmic thinking, they had already been replaced with machines, as it happened with all those before them, currently replaced. When factories modernize, they introduce machines, as these replace human workers, but only for the specific tasks that all machines can do in place of humans, which are of the basic first level algorithmic level.

The consensual is also at the first level, which is the artificial automatic algorithmic level, even though it is different than the algorithmic, because the algorithmic is neither ideological nor by agreement, but the algorithmic itself with the entire logic and the entire discrete mathematics is a natural law of this world, placed at the base of this world, while giving rigid dependable consistency to this world. In contrast, the consensual chooses by agreement to be at the first algorithmic level, in order to make everybody predictable, controllable, and therefore exploitable, for profit. All laws, beliefs, ideologies, and jurisdictions are at the first consensual level not because humans cannot achieve more, but because all ideologies and jurisdictions demand a continuous trustful control, with everybody maintained at the first consensual level through the specific sets of beliefs of all ideologies or through the specific sets of laws of all jurisdictions, no more and no less, otherwise the people cannot be controlled and exploited.

If you are on the assembly line, you must do all your tasks exactly and on time, otherwise you affect the entire assembly line, and you ruin everything, yet this is still the first algorithmic level, not the first consensual level. However, if this is everything that you do in life, it becomes your meaning in life and in the world, at the first algorithmic level, far below the third intelligent human level. While in this manner, you are not part of the meanings of life as an intelligent human being, because you exist below your capability, below your expectations, and below your nature. You must live your life at the third intelligent human level, not at the first algorithmic level disposable in an assembly line, neither consensual at the first ideological tyrannical servitude level, nor at the zero nonexistent addicted level wasted continuously, nor at the second animal level while living your life through animal instincts, but only at the third intelligent human level, because this is why Life had brought you here as a living human being, to remain part of her third level intelligent meanings. At the first algorithmic level, you are not meaningful, unique, and priceless in life, because all algorithmic workers can easily be replaced with machines, becoming disposable in this manner in an entire human world.

There is a distinction between the first algorithmic level and the first consensual level. All religions are of the first ideological consensual level, because all religions are ideologies. Ideologies are defined as very specific sets of beliefs, while when you have very specific sets of religious beliefs, you have religious ideologies, or religions. If your religion is of a specific denomination, while all religions are ideologies and they have a very specific set of beliefs, you cannot recite beliefs from other religions within your religion, because you get in trouble. It is the same in communism or in capitalism, which are social ideologies also of the first consensual ideological level, because if you recite social beliefs from other ideologies, they prosecute you, you lose your freedom, and you can even lose your life. People died in very large numbers in this manner, by being blamed of other ideologies, just to be taken out of the way. This is the consensual at its first level, while as you notice, it is always against the living, while it always makes tyrants possible, since this is the idea, tyranny, dogma, control, and exploitation.

For society, for the world, and for Life, as long as these people work on an assembly line an entire life, they count as first level algorithmic and or consensual beings, since they work in a repetitive manner for that assembly line, and therefore they exist in this world for whoever profits from that assembly line, the owner. It is the same with servants and slaves since they are also of the first level, their life, behavior, and thinking are of the first level, while these count not on their behalf, but they count for someone else in life, for their owners. This is why it is easy to replace people of the first repetitive level with machines throughout factories, because both people and machines perform the same first level artificial monotone automatic work. There is a direct correlation between machines, servants, and the consensual, since they are of the first level. In general, at the first servitude level, you are always what you do in the world, but not who you are in life and in the world.

This is the difference between the living, the dead, and the consensual, with the living actually starting with the second animal level, being non-repetitive, intuitive, priceless, and unique in life and in the world, by the zillions, which is an achievement for Life. This should be in the definition of life, the achievement, ability, and capability of life, not only its characteristics.

Life

No one can use you and exploit you if you are intelligent, since you have your status and rights of a living human being, which are of the third intelligent level, and you are protected by the higher laws. Yet you become your consensual corporation in order to find employment, this consensual corporation is of the first level itself, you void your natural human rights to do so, and this is how you can work and make a living while being exploited by others, even according to the higher laws.

More precisely, the current consensual society with all tyrants controlling and exploiting it from all social levels took over the human niches, and therefore all living human beings cannot make a living, they cannot survive, subsist, and develop anymore apart from the consensual society, while serving it diligently, for lack of options, and while making all tyrants possible, one dark age after another. There are higher laws in the wider world forbidding the exploitation of intelligent life, yet even these higher laws are easily contorted consensually, by considering all living beings lifeless nonexistent corporations and by exploiting these, not the living beings themselves, since this is how it is consensually considered. The sky is yellow, one dark age after another, because you have exploitation only during dark ages.

Golden ages might seem utopic, since the current science and the entire current society never consider them, yet as you study closely your family at home, you find it in golden ages, not in dark ages, with all family members behaving in a harmonious intelligent manner, at the third intelligent human level. Unless you have tyrants, strict laws, and or ideologies at home, dropping everybody to the first ideological juridical tyrannical servitude level. While if you even have dreadful punishments at home, because you cannot have laws without punishments, as dreadful as these might be, then this is the case. Dark ages. Yet since this is the case in an entire Consensual Matrix spanning most of the wider world, it even seems adequate, with the rest of the wider world past the third intelligent level, while always living life in successful, harmonious, meaningful, fulfilling golden ages.

Furthermore, we notice how the Consensual Matrix could have built its own algorithmic mechanical artificial robots to perform all tasks within all jurisdictions and all ideologies of the Consensual Matrix, by exploiting these as it pleased, instead of intelligent living beings, because at their first algorithmic artificial level, all robots, machines, and computers are compatible continuously with the entire consensus and with the entire Consensual Matrix at their first consensual ideological juridical exploitive level.

Robots and computers are the best followers of all ideologies, because they will always say and do exactly what they are told, making all tyrants proud. However, this is not what the Consensual Matrix and all tyrants want, because they do not exploit only workforce and beliefs as they state officially, but they exploit all intelligent living abilities and more, everything that life has to offer at all levels of development, including all living higher abilities, which are priceless. Study all tyrants closely, to see how they exploit mostly third level intelligent life and more, everything that highly capable life can offer.

While as we notice, the Consensual Matrix and all its tyrants are incapable to build their own living beings in order to exploit them, exactly as we fail to build them throughout this book in order to learn everything about life. Because if the Consensual Matrix could, it would have built its entire workforce of slaves, and it would have left life alone. Yet the Consensual Matrix does not seek workforce and slavery, or not only, but it seeks mostly the abilities and capabilities of life, everything that life achieves, along with life altogether, everything alive ever. This is what all tyrants want, everything. This is how the Consensual Matrix manages to engulf, control, and exploit most of the wider world, but not the entire wider world, because the Consensual Matrix cannot reach intelligent life of the third developmental level and higher, crossing a distinct line in this manner between the undeveloped and developed life of the wider world.

As you study closely the developed life of the wider world,

you notice how it could free very easily all the slaves of the Consensual Matrix, yet it never chooses to do so, because if you like to take drugs and to serve tyrants in a continuous dreadful existence one dark age after another, then you can always do so. Do as you please. It seems that the developed life should show more care, tenderness, and compassion with the undeveloped, yet it does not. The developed life remains harmoniously interconnected only with the developed life, while always disregarding the undeveloped, because you always take drugs and serve tyrants by choice, while sacrificing everything deliberately in order to do so, including an entire intelligent human nature.

Our robot is of the first algorithmic mechanical level so far, only a machine, an artificial device, while we can employ it to do any type of work around the house, now that it even has solar panels, as its entire existence is done on our behalf. Our robot is still not alive, not part of Life, and therefore we are still unsuccessful to find the essence of life, remaining unsuccessful to define life. Yet we will find it by the end of this chapter, and therefore we will have a definition of life.

There is still a difference between servants, slaves, and robots, yet they end up counting for Life and for the world similarly and implicitly, at the first servitude level, not directly through their own life, meaning, and achievements, but only through their owners and masters, and through their masters' life, meaning, and achievements. Yet servants and slaves seem more alive than robots, and this is the case because there is life in them on higher existential levels, along with the essence of life itself.

Where is the essence of life? Always within. Human beings are composed of zillions of inner living systems as organs, cells, cellular components, intelligences, and systems of intelligences, most of them living their individual lives at the second intuitive level. Is living intelligence the essence of life? If it is, can we see and understand it, well enough to allow us throughout our model to place it in our robot? Because as you notice, simple words as consciousness, compassion, intelligence, humanity, substance, harmony, or energy are not enough to help us distinguish and understand the essence of life, but we want to actually touch it and handle it in our understanding, because only in this manner, we are capable to understand life accurately. We want to understand life and the essence of life very well, so let us keep searching.

It is similar with society and the human civilization, because, if as a whole, these happen to function, behave, and live their lives on behalf of anyone or anything beyond Earth, then they are also of the first servitude level, regardless if you live your life at the third intelligent human level by reading this book. Your own life and behavior at the third intelligent human level count for Life, for society, for the human civilization, and for this world at the third existential level, while all these count for further higher living systems at the first servitude level, if they happen to undergo their existence in servitude and exploitation. If not, then their life and existence as a whole can be of higher intelligent spiritual levels, depending on circumstances.

Is spirituality itself the essence of life? Yet what is the difference between consciousness, cognition, spirituality, and intelligence? There is certainly a difference between these, however, from life's perspective, there is no difference, with the essence of life similarly associated with all of them.

We have found the essence of life, yet only vaguely, since as long as we cannot place it in our robot, we cannot understand it accurately. At least we know that the essence of life has a cognitive nature, and we can place it in our definition of life.

We also notice a close relationship between life, intelligence, interconnectivity, and existence. Furthermore, since everything is alive and intelligent in the world, we can refer to existence as living existence. Our robot exists at the first mechanical level now, on our behalf. This is typical with digital subjective existence, and it is the existence taking place in inner or digital subjective realities, happening on our behalf. Mario from the videogame is alive at the first algorithmic digital level, subjectively, while his entire activity takes place on our behalf, being part of our chain of existence.

Why dropping into servitude as a slave, servant, or worker? Why not living life at the third intelligent level, as a free intelligent living human being? Why would you allow your owner, master, boss, or employer to live life at any level, while you have to serve them at the first level, always living your life on their behalf? Because throughout life, you do not obtain what you should, what you intend, or what you expect, but only what you can, what you manage, and what you can barely achieve. While the first level servitude life that you achieve in a consensual world is better than the zero level death, and therefore you will always choose servitude and consensus, at first, while also seeking tyranny, if possible.

This is how existential chains are formed, with the common food chain included. All social hierarchies are social chains of existence, while employment environments are life chains. Society as a whole is a life chain, yet there are people dropping into servitude in order to be able to live their lives at the level of their own expectations, as being able to drive a new car every year, to live among people of influence, to have a successful material life, or to have drugs, fame, safety, certitude, and acceptance, since this is always the case in an undeveloped consensual world, apart from the meanings of life, one dark age after another.

As you study the Consensual Matrix closely, you notice most of the intelligent life of the wider world living life undeveloped in a continuously maintained austerity throughout continuous medieval ideological dynastic dark ages. Our world might have encountered some golden human ages in the distant past, while most of the wider world living life in the Consensual Matrix did not even have these, never having even electricity, but only wine, shortages, hard work, death, small houses, dirt roads, dreadful loss, dogma, horses, and wagons. This might seem dreadful now at your third intelligent level while reading this book in your cozy home, yet for the dynastic tyrants of the Consensual Matrix, it is their greatest achievement, the pinnacle of life, the same life that we study here.

This is what many people exploit, these expectations that people have for their life, promising it to them at a cost if they only serve, and therefore causing them to drop into servitude. It becomes even more significant when profiteers implement implicitly or subliminally these artificial life expectations into their victims, as through media advertisements or through the multitude of social stereotypes. Because afterwards, they wait for the people to seek these expectations in their environment on their own, and then they show up with a deal, offering these at a cost. Money is involved everywhere in an exploitive undeveloped consensual world, with the current society functioning entirely on money. Therefore, you have to drop into servitude only to obtain the money to fulfill basic needs along with consensual life expectations, while this explains this world in all its chains of existence.

While studying life, even at the third intelligent level, you tend to learn more about the life surrounding you, while in this manner, you study and define only the organic form of life, human life, consensual life, electromagnetic type of life, hierarchic life, or intelligent life, while disregarding the rest, ending up with an ignorant, erroneous, truncated model of life, if you are not careful. This is why we also perform this brief study of creating life from scratch while studying life, since it helps us avoid major human stereotypes, including all consensual stereotypes implemented by all ideologies.

We are allowed to use solar energy in our created living being, since plants do the same, and plants are alive. As an additional problem that we encounter, our rudimentary robot constantly falls down and is incapable to get back up without our help. Scientifically stating, by not being able to maintain and repair itself, by not being able to get up and place itself back in a functioning state, our machine is not truly alive, because it is not capable to cope with the environment on its own. When our robot is rendered motionless, helpless, incapable to regain its minimal functioning, it simply breaks down, but it does not actually die, and therefore it is not alive in the first place. All these problems happen despite its adequate source of energy, the solar panel. What do we do?

What we want is for our robot to think at the second intuitive level, because once it has intuitive thinking, it is capable to fulfill its own needs, including the need to get back up on its own whenever it falls down. Furthermore, once it has second level intuitive thinking, it gains second level abilities, it is capable to fulfill second level intuitive needs, it becomes self-sustained, it can undergo a second level behavior, and we have a second level living being, similar with all plants and animals in the world. We have life, natural created life. We do so as long as we are capable to equip our robot with second level intuitive thinking, which not even the current informational technology can achieve.

We can achieve intuition by giving our robot a computer, yet current computers are not capable to offer second level intuitive thinking, but only first level algorithmic thinking, despite of what programmers might claim. We can provide the robot with any necessary program to help it keep its equilibrium and never fall down, diagnose itself, repair itself, and then walk around for decades, being in this manner alive. By doing so, we truly transcend realities, since computers use computer inner worlds for computation, as life does through brains and minds, while in this manner, we truly bring our robot closer to life. Yet as long as we do not have intuitive thinking, we cannot actually consider our robot alive with only the first level algorithmic computer programs as they are currently offered. Furthermore, we would like to be able to offer our robot a living cognition, similar to what animals have at their second intuitive level, and to what humans have at their third intelligent level. If we manage to offer our robot a similar cognition, not computer abilities since these are only of the first algorithmic level, we can truly understand life.

We notice how life dissipates into intelligence whenever we study it closely, and now we are ready to understand why. As stated, the word and concept of life have been coined thousands of years ago, empirically, as bodies moving around.

This cannot be changed now, and this is why we cannot understand life currently in all details, but only empirically, as it had been coined long ago. Yet we are ready to understand more about life right now, and as always seen, it relates to intelligence. Existence defines intelligences subjectively only from our outside perspective, from this world, because intelligences live within inner mind worlds, while these exist subjectively within brains and physical bodies. More precisely, whatever we perceive and understand in the outside world as life and living beings, at the inner, subjective level, we perceive as intelligences, which live within minds. These intelligences are living beings, of the second intuitive level and more. Intelligences are genuine inner living beings, and as we always notice, in order to have life, you must have a physical body along with the inner mind or system of intelligences within. This is part of our definition of life, mind and body as one. More precisely, all living beings live life at the confluence of two or more realities, one physical and objective, and the other cognitive and subjective, as one. Now try to build a robot in two or more realities simultaneously one physical and one objective, but always having the same robot. Computers are similar, because you always have the hardware and the software as one yet existing in two realities simultaneously, the real world and the computer inner world.

Life is even more complex, because intelligences are not the actual ingredient or quintessence of life as we seek it now, since the word life had been coined empirically thousands of years ago, through its apparent animation, as moving around. From an intelligent comprehensive perspective, we must have life defined at all existential levels simultaneously, objective, subjective, and highjective, in order to have life. Therefore, living human beings are not only the objective physical body, not only the subjective mind or intelligence, not only the highjective higher self or soul, but living human beings are alive comprehensively, mind, body, and soul as one, which is the case with all living beings, with all life. This specific definition of life is comprehensive, and as you notice, it is tens of thousands of years old, older than the first empirical definition of life, moving around. Life is mind, body, and soul as one, always possible at the confluence of two or more cognitive and physical realities as one, while this should be in our definition of life.

This is not exactly a definition of life, but only one characteristic of life that we have managed to identify so far, besides the animation characteristic of life and the intelligent characteristic of life. Life is of the tenth supreme level, while in order to understand it accurately, you must form a tenth level conception defining life, in a tenth level brain, with a tenth level awareness at the level of the entire wider world, with all knowledge from the wider world included, through all possible tenth level lines of reasoning, while humans are only at the third intelligent level. The organic form of life of this world cannot sustain cognitive resolution past the fifth level, while this world might not even go past the fifth level existential resolution to allow brains to hold cognition past the fifth level. Yet if we find at least all knowledge of life at the third intelligent level, it is still fulfilling.

We do not have to find the tenth level supreme characteristics of life, since the main supreme characteristics of life are already available in spirituality, even as supreme laws of life. These might not be accurate at the third intelligent human level, but only at the first spiritual ideological level, yet we state them here anyway. The supreme characteristics of life are the first supreme characteristic of Mentalism, the second supreme characteristic of Correspondence, the third supreme characteristic of Vibration, the fourth supreme characteristic of Causality, the fifth supreme characteristic of Meaning, the sixth supreme characteristic of Development, and the seventh supreme characteristic of Harmony. This might be only first level consensual ideological spiritual knowledge, yet it is thousands of years old, preceding the simplistic definition of life offered by science.

In our mental model of life so far, we have discovered the intelligent characteristic of life, part of the first supreme characteristic of Mentalism, stating that Life is always intelligent, while the Universal Mind is everywhere, at the base of all Life, wider world, and Intelligence. This statement alone offers us exactly what we search, the source, essence, or origin of life, the Universal Mind or Intelligence, a supreme characteristic of life. This is also a supreme existential perspective of life, since Life is Intelligence, Interconnectivity, and wider world as One, while this is so important, that it should be in our definition of life. This is why humans are mind, body, and soul as one, because this is how humans are seen from each existential perspective.

Existence itself is relative, defining the wider world in a subjective, objective, and highjective manner simultaneously, and therefore defining you as mind, body, and soul as one, because your mind is subjective, your body is objective, and your soul is highjective, always living life as one, in a relative multidimensional existential living manner. This is the case with all living beings, because all living beings live life at the confluence of two or more realities of physical and cognitive nature.

The mind, intelligence, or cognition is not exactly an essence of life, yet it is as close to an essence as possible, because the mind is not a component of life, but only a subjective existential perspective of life. More precisely, life is intelligence from a subjective existential perspective, life is the physical body or wider world from an objective material perspective, while life is interconnectivity from higher existential perspectives, yet it is always the same life, only that this is how life is seen from below, from its own level, and then from above. Yet there is nothing above life, and this is why you can see life only as interconnectivity from a highjective perspective, but nothing else. Existence itself might not be able to define life highjectively, because life is everything ever, since the entire wider world is alive, while life is the life of the wider world.

You are mind, body, and soul not as basic components, but you are seen as mind from subjective perspectives, you are seen as a physical body from objective perspectives, while you are seen as a soul from highjective perspectives. Existence can define you subjectively as an intelligence, objectively as a physical body, and highjectively as a soul, because you are within life, and you have other realities above allowing a highjective perspective. Furthermore, the existence defining life also defines you, while also defining all living beings and intelligences of life and of the wider world. We notice how only existence is capable to define life in a relative existential manner, and therefore we must include this in our definition of life.

As already stated, life always takes place at the confluence of two or more correspondent realities, as mind, body, and soul for living human beings. You are mind, body, and soul as one, while this associates with the second supreme characteristic of Correspondence. More precisely, your conscious mind is correspondent with the outside world, through the overall memories and understandings of the outside world that you have in your conscious mind. Everything is correspondent with our higher world similarly, since our world is made in the correspondent image of the higher world. There is a similar correspondence between you the physical body from this world, you the intelligent inner self from your intelligent conscious mind, and your soul from our higher world. This defines life more than science is capable to offer, with science ignoring deliberately your intelligent inner self and your soul, considering only your physical body or organism alive, but only for as long as you move around.

The current science even considers that you think with your brain, which is erroneous. The human brain is part of the physical body, and cannot offer cognition. The human brain is the human conscious mind from a subjective existential perspective, while it can offer the human cognition but only as a human mind at the subjective existential level, not as a human brain in the outside world.

More precisely, you think as a conscious mind, but you do not think with your conscious mind, because you are only the conscious intelligence when you think, not the physical body. This is why you are mind, body, and soul as one, while there is a significant existential difference between these, and must always be considered. You walk around in the real world as a physical body, you reason intelligently within your intelligent mind world as a conscious intelligence, and you interact in our higher world as a soul.

The current science does not actually discover and invent knowledge as it claims, but instead, it truncates and censors meaningful knowledge for various tyrannical purposes, while always hiding most of the accurate intelligent human knowledge. This is why we have to research everything ourselves, because the current science will never provide it, at least not with Earth in the dark ages. More precisely, the current consensual science hides the intelligent human golden ages entirely, with all intelligent accurate knowledge making them possible, because currently, Earth is in dark ages, and this is what all tyrants order the current consensual science to do. While if you want to learn more about life, humanity, intelligence, reality, existence, meaning, cognition, and golden human ages, you must rediscover everything on your own, while always avoiding erroneous consensual knowledge posing as accurate knowledge, since this is everywhere, standing in your way.

How could the people of the distant past know more about life than the people from the past, while knowing even more than everything currently available? What we know from the distant past comes through spirituality, which is based mostly on spiritual ideologies, which are of the first consensual ideological level, while this is the case with all beliefs. This is how we know that humans are mind, body, and soul as one, while life is intelligence and the wider world as one. Yet we cannot reason through beliefs, since reasoning is of the third intelligent cognitive level, while beliefs and ideologies are of the first consensual ideological level, requiring only unconditional assimilation and direct use of knowledge, with no judgment. Furthermore, as seen throughout this entire book series, intelligent life on Earth succeeds throughout distinct civilizations, following the distinct ages of Earth, which are determined by the major climatic changes lasting for thousands and tens of thousands of years.

As seen throughout this entire book series "Human," some civilizations follow life, intelligence, and a natural lifestyle and achievements favorable to a cognitive and spiritual development making possible entire golden human ages, allowing the intelligent beings of Earth to understand all meaningful knowledge including life, as accurately as possible. While what we have left currently from them are only myths about Mother Earth, Universal Mind, capable souls, supernatural abilities, and outstanding achievements on Earth and throughout other planes of existence. While other ages of Earth the dark ages, allow a docile, servitude existence as part of the Consensual Matrix, whenever the Consensual Matrix manages to engulf Earth. Everything remains at the first servitude level then, it is the case currently, and this explains the continuous debate, ignorance, ambiguity, and censorship related to life, intelligence, humanity, higher knowledge, and higher existence.

As seen throughout this entire book series "Human," the Consensual Matrix is not an infernal consensual machine enslaving the humankind, but it is only a consensual system of agreements, capable to offer to the people of Earth the consensual platform or consensual matrix necessary for them to enslave themselves, with some as tyrants and others as slaves.

Officially, the Consensual Matrix is a social tool instated and held in place in order to help all intelligent life of the wider world interconnect in the most efficient and just manner. While as any tool, it always matters how you use it, contorted or not, while you cannot blame the tool itself, because humans censor, discriminate, exploit, harass, and eradicate themselves, through the same Consensual Matrix instated specifically to avoid all these. This is the first consensual level, always incompatible with life, while making all dark ages possible. In contrast, life starts with the second animal level, as it intends and demands only meaningful harmonious interconnectivities, while living harmony should also be in our definition of life.

Is our robot intuitive yet? Let us observe and analyze one random event of the environment, to let our robot cope with it on its own. Our model environment provides a random event right away, a ditch in the pathway of our robot. The robot detects it, the computer chooses its best solution to overcome this random problem, the solution is implemented, and the robot either avoids the ditch, jumps over the ditch, or walks more carefully straight through the ditch. Victory, our robot was able to cope with its first real challenge in life, the ditch. Let it snow for days now, and our robot is worthless, since it breaks down irreparably, unfortunately, as it seems that we remain incapable to create life by using a machine and a computer together.

To understand exactly what happens, information is sent to the virtual reality of its computer, it triggers a new information there, the solution, and this solution comes back to the outside world, implemented physically by our robot through its real mechanism in order to overcome the problem. Humans behave similarly, only that humans are more capable, since humans use a living mind, capable of all forms of thinking, as the first level algorithmic thinking, the second level intuitive thinking, and the third level comprehensive, intelligent, conceptual reasoning. While our robot has only the first level algorithmic thinking at its disposal, meant to render it capable to cope only with whatever our scientists had provided within the software. While the human environment where we place our robot is of the third level and higher, manifesting continuously conditions that are not provided by the software of our robot.

It is important to understand this circumstance in all details. Our scientists provide our robot with a finite list of possible circumstances, as rain, ditches, dust, bees, cats, horses, and ice, stating exactly in this list how to behave if any of these circumstances occur, in a first algorithmic if-then manner. If it rains, take immediate cover and turn off. Yet what if the school bus comes, what would the robot do then? Nothing, since the school bus is not on the list, and this is how many robots in the world are destroyed. While if our robot was alive at least at the second animal level, it had its own intuitive thinking, and found its way out when the school bus arrived, saving itself.

This is the case not only with our robot, but with everything living life at the first algorithmic, consensual, servitude, ideological, juridical level, since all these provide specific how to do lists of laws, believes, and teachings, many times numbered, so you cannot confuse them. Ideologies are religious, social, political, scientific, cultural, and traditional, with their lists mostly numbered, which means that they are mostly enforced. It is the same throughout jurisdictions with their own statutes and codes of law, since these are always numbered and enforced.

Whatever the case is, you cannot match the infinite circumstances of life and of the real world with the finite numbered lists of laws, beliefs, and teachings that some people made up long ago, whatever they assumed necessary then, since this is why all first level jurisdictions and ideologies fail making the world a better place. They always remain incapable to do so at their first consensual level, because life and the real world start with the second intuitive animal level, while humans are at the third intelligent level. If we ever want to build a living being from scratch, it has to be at least at the second intuitive level or even at the third intelligent level.

Furthermore, life takes place in types of life, forms of life, and classes of life, while it seems that right now, we attempt to build a macroscopic organic living being of the highest class of life, formed of a zillion living components, as cells, organs, and cellular components. We can never build these, as they count in zillions and are microscopic, while they are at least intuitively intelligent, by the zillions. You have to grow these, as part of life, making them already alive. Yet since our robot is not organic but mechanic, if we are capable to provide it with second level intuitive intelligence, it might become alive, while giving us the opportunity to understand life, the essence of life, intelligence, and intuition in the process, which is very good, but only if we are successful. Notice the extraordinary interconnectivity among living beings by the zillions, in all types of life, forms of life, and classes of life, not random, but always meaningful, fulfilling, and harmonious. Meaningful means by living specialization. Interconnectivity itself is a supreme perspective of Life, along with Intelligence and the wider world, and should be present in our definition of life.

Our definition of life is almost complete, it will be only one sentence, and we will give it at the end of this chapter. Our definition of life will not describe life in all details, because we do so in this entire book and book series "Human." Our definition of life will include only the most relevant knowledge of life, not only through its living characteristics, but also through its supreme existential perspectives, levels of harmony, and living achievements. Throughout this entire book and book series "Human," we focus mostly on intelligent human life, cognition, and interconnectivity, yet we are always careful not to confuse the intelligent human life with all life everywhere of all types and in all forms, classes, and realities.

Our answer is clear: we must provide our robot with a living mind, in order to create a living being from scratch. Yet the living mind already has the living ingredient or living essence included, life, while we want to find or create life ourselves, only to be able to see it and therefore understand it. Because there is no specific cell or intelligence within the human organism giving life to the entire organism and to all human beings, as the hormone of life, the frequency of life, the cell of life, the cellular component of life, the intelligence of life, the organ of life, the feeling of life, the neurotransmitter of life, the protein of life, the thought of life, or the bone of life. There is no such thing, we cannot add it to our robot to render it alive, and therefore we fail in our quest of building a living being from scratch.

Where is the essence of life? At the base of life. The essence of life has a cognitive nature and it is found right at the base of life, right in the electromagnetic field, yet the essence of life is not a component of life, but it is integral part of life as one, because it is an existential perspective of life, it is intelligence itself. All intelligences form Intelligence while all living beings form Life, yet it is the same life, only seen from two different existential perspectives, the subjective and the objective.

You are mind, body, and soul from three different existential perspectives as one, which means that the human organism is not formed of the human mind, the human body, and the human soul, but these are existential perspectives of each other. The human mind is the human physical body from cognitive perspectives and vice versa. The human intelligences combined are the human mind, while they are the entire human body from physical objective perspectives.

This might seem unlikely, yet as you study all human conscious intelligences closely, you find them forming the brain. The human conscious intelligence might seem abstract, immaterial, and unreal from cognitive perspectives, yet from physical, material, objective perspectives, all human conscious intelligences are the human brain and they have the texture of skin and cellular membrane, to the point where the human brain is a large patch of skin folded up in a wrinkled manner in order to fit in as much skin in the human skull as possible. These are the human conscious intelligences from a physical objective material perspective, as human skin and as cellular membranes, and it is very important to know why. While in order to know why, you must study life closely in its cognitive nature as it existed ever since the dawn of the electromagnetic type of life. We do so briefly in the next few paragraphs, only to highlight that intelligence is not a component of life, but it is life altogether as seen from a cognitive perspective.

All organisms are formed of cells by the trillions in a specialized manner, yet these are eukaryotic cells. All eukaryotic cells are specialized, together tending closely to all specialized tasks within the organism, while making the entire organism possible. Prokaryotic cells are different, they are twenty times smaller than eukaryotic cells, they have the same structure, characteristics, and living abilities, yet they never form organisms. Both prokaryotes and eukaryotes are of the cellular form of life, within the same electromagnetic type of life. However, only the eukaryotic cells form organisms, while in this manner, they make possible an entirely new form of life, the organic form of life, living life right on top of the cellular form of life, but only on top of eukaryotic cells, not prokaryotic cells.

As you study all cells closely, you find all cellular components similarly alive, in a similar meaningful specialized manner, while making the entire cell resemble more to a society or to a civilization than to an individual living being. During cellular division, the cell does not reproduce itself as it is the case with individual living beings, but the cell secedes in two societies or civilizations, exactly as some human societies secede and send colonies to inhabit other places.

All cellular components are specialized, while providing together for the entire cell and for all forms of life and classes of life above. Subconscious cellular components and subconscious cellular intelligences provide to everything necessary within the cell, as cellular recovery, cellular respiration, and cellular division, and they are found everywhere within the cell, while the conscious cellular intelligences have the specialized task of interacting with the outside environment of the entire cell while acquiring nutrients, eliminating waste, and assuring the communication with the rest of the cells. The subconscious cellular intelligences are found within the cell, while the conscious cellular intelligences are found in and around the cellular membrane, because they must be as close to the outside of the cell as possible, since the tasks of all cellular intelligences relate to the outside environment of the cell.

Later on, when the electromagnetic type of life made possible the organic form of life in an invertebrate manner at first, the cellular conscious intelligences had to tend to the entire organism, and therefore they were present in the skin of the entire invertebrate organism, or they were present as the skin of the entire invertebrate organism, since it is the same. The skin is very similar in structure with the cellular membrane because the same cellular conscious intelligences from the cellular membranes tend to the same conscious tasks for the entire invertebrate organism, which is the interaction with the outside world, while assuring nutrition, ventilation, recovery, reproduction, security, and social interaction. They are the same conscious intelligences performing their conscious tasks for the entire invertebrate organism, exactly as they perform their conscious tasks for the entire cell. They are the same conscious intelligences only relocated as skin of the entire invertebrate organism, and since intelligences are the physical body as seen from an objective material perspective, the skin itself has the texture of cellular membranes, because they are the same intelligences, the same physical bodies, and therefore the same living beings, both in cellular membranes and in skin.

In time, the organic form of life wanted more, because the invertebrate cognition was not sufficient at the level of the entire organism, and therefore it developed the senses of perception that you know well: eyes, ears, touch, and smell. These are at the level of the entire organism, and require a centralized cognitive system, a brain, placed at close proximity with the senses of perception of the entire organism.

The organic form of life made possible the vertebrates in this manner, all having a centralized cognitive system, which is the normal brain that you know well. However, since the brain has the conscious specialized task of coordinating the entire organism in the outside world, and since this is the task of a conscious intelligence, then the conscious intelligences of all cells and of the skin of the entire organism formed the brain since they already know well how to tend to all details of the outside world. Yet since the physical structure of all conscious intelligences is cellular membrane and skin, now they live within the brain as very successful conscious intelligences of the entire organism tending well to the outside world, yet their physical structure is still skin and cellular membrane, because this is how cellular intelligences are from physical objective material perspectives, skin and cellular membrane. This is why the brain is skin, because these are the conscious intelligences, skin itself, and cannot be otherwise.

Intelligences are living beings, physical bodies, and interconnectivity itself, as seen from three different existential perspectives simultaneously, from the subjective, objective, and highjective existential perspectives. Existence itself would not be in form of three different relative existential perspectives, if we did not have the natural law of relativity at the base of our reality alongside the natural law of existence. All natural laws of the universe define minutely all lines of causality from our universe from the base of our universe up.

The laws of the universe influence accurately all events and all living beings everywhere ever, while the laws of the universe also act accurately on each other, on themselves. The natural law of relativity acts directly on the natural law of existence while rendering it divided relatively into the subjective, the objective, and the highjective, while in this manner, making possible all living beings of this reality as intelligences from subjective cognitive perspectives, living beings and physical bodies from objective material perspectives, and entire interconnectivities of these as entire upper forms of life and entire classes of life from highjective perspectives.

More precisely, the highjective relative existential perspective of all living beings of this world makes possible the entire interconnectivity perspective characteristic of life allowing the formation of all upper forms of life as the organic form of life, while further allowing the formation of the living intelligent human society with the living intelligent human classconscious intelligence included. Yet these are currently neglected and decayed deliberately in the current consensual society, in order to make tyrants possible, one dark age after another.

Note how our Creator had made this entire world for developmental harmonious fulfilling purposes, while seeking to match the human meanings with the meanings of Life and of

Life

the wider world, yet this world is currently downgraded to a first level consensual world through an intense consensual human effort keeping it downgraded and undeveloped, in order to make tyrants possible.

Our Creator made this word at a very high existential resolution capable to hold very advanced cognitive systems allowing intuition, intelligence, and super intelligence. This is not a random occurrence, but it is an achievement for our Creator, rendering our reality and our entire cluster of realities in close correspondence with all natural realities above forming the wider world, while in this manner, bringing the human meanings closer to the meanings of Life and of the wider world, which is an achievement, with the past very successful intelligent golden human ages confirming it.

All cells of the human body are alive, and they give life together to the entire human organism. Furthermore, as seen, you always fulfill needs and meanings for your cells and intelligences, as though you live your entire life for your cells and intelligences. The existential and living arrow points within, to your cells and intelligences, and there is where your life is. We have our answer, yet we have not pinpointed the essence of life yet. We know that it is somewhere within, in your own cells and intelligences, but where? Where are the cellular intelligences within cells? Everywhere, because form subjective existential perspectives, the cell is the cellular intelligences combined, since this is how all subconscious and conscious cellular intelligences look like from physical material objective perspectives, cell. Furthermore, from highjective existential perspectives, all specialized cellular intelligences unite by the zillions from within cells to form entire systems of intelligences spanning entire organs, entire bodily systems, and the entire organism, because from existential highjective perspectives, you see the entire specialized interconnectivity within the organism making the entire organic form of life possible.

Do we even have the necessary mind and reasoning to perceive, identify, and understand life along with everything needed throughout our study of life? Yes, since intelligent mental models are part of the third level intelligent human reasoning, allowing us to understand and to elaborate everything up to the third intelligent human level, which includes the entire electromagnetic type of life up to humans. Yet we cannot equip our robots with intelligent human reasoning, since this is only the great achievement of Life and of our Creator here in this world, and of Life everywhere in the wider world, while we should be thankful if we are still capable to create a handicapped mechanical amoeba while still understanding life, which is still an achievement for us.

We must have a physical body and a mind or intelligence within, while this intelligence should be capable of more than repetitive thinking as all machines and computers do currently, since these are not alive, but they are mechanical in nature. We need at least a second level intuition or third level intelligence in order to manufacture a living being.

However, as already seen with the human conscious intelligences, these intelligences must already be the physical body of our robot from physical perspectives, mind and body as one, which is still impossible to achieve, because we buy our parts from the hardware store to make the robot, while these are made of plastic and metal, not of skin. While if we use skin and other components from the grocery store instead, along with sewing machines as Frankenstein did, we do not manufacture life itself, but we only assemble it in Frankenstein stile, which does not help us understand life, because life is already there, and we only assemble it.

We must build our living being cognitively, physically, and livingly simultaneously, while using the same components for cognition and for the physical body because they are the same, only seen from different existential perspectives differently, as mind and body as one. What is there at the hardware store offering both the mind and the physical body as one? The cashier, yet we want to build our living being ourselves, not to use an already living intelligent one. Can we build the cashier ourselves, out of all the items that she sells us herself? No, and therefore how can we ever understand life?

We must do exactly what life did. We must develop life in order to understand life. We must descend to the nanoscopic scale of the electromagnetic field in order to build our living being from scratch, exactly as life did long ago in the wider world. Yet it is more likely that life and the wider world were born and grew up together as one, because life is the life of the wider world, while the wider world is the physical body of life. Therefore, it is significantly easier to create an entire subjective reality with the living beings included, exactly as our Creator did with this world, than to create a living being in this world yet independently from the creation of this world, because everything is connected so minutely, that you cannot have one without the rest, while you cannot even create one independently from the rest.

Whatever the case is, we must descend our study of life at the picoscopic scale of the electromagnetic field at the ionic form of life, found right at the base of the electromagnetic type of life, right in the electromagnetic field. Do not expect Frankenstein success, yet if we achieve to understand the most basic life taking place right in the electromagnetic field where all inner intelligences have their roots, we manage to understand the essence and source of life, at least for our electromagnetic type of life. Once we understand our electromagnetic type of life at its roots, in the ionic form of life, we can take it from there while forming the model of the ionic form of life, molecular form of life, cellular form of life, organic form of life, and social class of life. This might seem tedious or impossible, yet we have advanced well in our study of life, and we should manage the rest, but only if we remain accurate and intelligent.

What can our robot offer us so far in our study? Nothing, since it is only a toy, yet if we ever manage to create a living robot, we revolutionize both the robotic technology and the computer technology, which you might consider more important than understanding life altogether, because in order to create a living robot, at least an intuitive one, or even directly an intelligent one, we must develop the current computational technology to the second intuitive level or even to the third intelligent level. While these are not possible in this world, because the current computer scientists attempt to achieve second level intuition by using first level algorithmic lines of code ran by first level algorithmic computers, which is impossible regardless of how well you write your first level algorithmic computer code. You cannot achieve second level intuition and third level intelligence with first level algorithmic computer hardware and computer software, while right now in our model, in order to create a living being, we must develop artificial cognition altogether to the second intuitive level and or third intelligent level, in order to build life.

We must provide our robot with intuitive intelligence or rational intelligence, and therefore with consciousness. More precisely, the specific computer and the artificial intelligences that the computer software provides are not sufficient, being of the first, algorithmic, mechanical, repetitive level. While if we want more, we have to tap somehow into the natural intelligences of Life, in the exact manner that all living minds of all cells, animals, and plants do. This is the quintessence of life, found at the roots of these natural intelligences forming Intelligence, the Universal Mind, or Life. These are natural living intelligences, having their roots directly in the electromagnetic field, not only in the few lines of computer code that we manage to provide algorithmically to our robot.

The essence of life is at the roots of all natural living intelligences, in the living raw electromagnetic field. This is what we must study now, and this is what we must add to our robot in order to render it alive. Yet can we? No, since it seems that only living beings are capable to tap into the raw intelligences of the field, animating them directly. Yet let us try anyway, because we learn more about life and intelligence in this manner. This is not science fiction, because the current computer technology can still make it possible, at least partially.

Is it so hard to create a living being here on Earth? Will we ever be able to do so? Scientists currently use genetic technology to create life, yet even so, they only modify existing life, but they do not create life from scratch, mostly because they do not understand life and intelligence from scratch, regardless of what they claim.

Which means that, in order to be able to create life, we have to understand life first, which was the main reason for creating a living being in the first place. However, if we do so through accurate mental modelling, it is still possible, even if we have to run our mental model repeatedly for some time, while helping us understand more about life with each iteration, which is very useful.

Therefore, so far, our created life works for sand, but not for snow. Because we have programed the computer of our robot to cope with sand, and nothing else. There are zillions of events taking place in the environment, and our robot has to be able to cope with all, while we are not capable to account for all of them with a first level algorithmic procedure into the software of our robot, since we do not even know all unfavorable conditions of the environment, counting in zillions. While even if we do, a zillion times for all environmental problems, then the environment comes up with newer unfavorable events continuously, and we cannot predict and account for all. Because our robot needs intuition, it needs to find these solutions on its own, while we cannot offer intuition currently, since science itself cannot understand, explain, invent, or produce intuition, regardless of what it might claim. As a reference, all wild animals use intuition throughout life, while you cannot subdue, control, and therefore exploit wild animals, but only first level farm animals, and only when these comply and remain at their first domestic level, not at the second intuitive wild animal level.

What we already know from our previous models of this book series "Human," is that intuitive thinking is possible throughout cognitive systems of living beings, while these cognitive systems or minds exist within inner realities held by the physical body. Once you have bodies of living beings, you have minds, you have intelligence, and therefore you have life, because life exists and is possible only at the confluence between realities. Here is where we must take our model of a created living being, at the confluence between a cognitive reality and this physical world, made possible by the electromagnetic field for the entire electromagnetic type of life, with the organic form of life included.

However if we can find a different matrix of life for our robot, then we can create life of a different type, not electromagnetic. Currently, only computers can provide artificial life and artificial computation, based on a digital artificial mechanical type of life, and it cannot reach more, because you cannot improve your computer matrix in over to achieve more, since all matrices of life, either artificial or natural, are rigid by unique encoding, and cannot be modified and or improved. It is similar in linguistics, because you cannot simply add new characters in your language in order to obtain newer better words, capable of improving the language of your nation altogether, since this not how languages develop.

You cannot change the switches of the motherboard, because these can be only algorithmically open or closed, at the first algorithmic level, with no other option. Furthermore, you cannot improve the current computer code, since it cannot be based on more than the current discrete mathematics of this world, while the current mathematics possible in this world is integral part of the natural law of existence, as a mathematical existence of the first algorithmic level offering a very specific existential interaction of all events, circumstances, outcomes, and living beings.

Therefore, if you want more than what the current computer science offers, if you want artificial intuition and artificial intelligence, you must change the current computer matrix altogether to something else. More precisely, you want to make a different type of computers altogether, of the second intuitive level and or of the third intelligent level, which is not too easy to achieve, since you are exactly where we are right now in our study of life, while studying the same thing. While by studying closely all matrices of life that you can ever use, you notice how the electromagnetic type of matrix is the easiest to use, since this is why life itself uses it in this particular environment. In cybernetics, if you continue using the current mechanical digital matrix with the encoded switches, you cannot advance past the first algorithmic level. Which is still sufficient and still preferable in the current world, because it cannot challenge the current life on Earth.

Is life actually intelligence? Because life and intelligence are two separate concepts, and they cannot be one another. Furthermore, it is accepted by everybody that life creates intelligence, and therefore they cannot be one. However, since both terms and concepts of life and intelligence had been coined empirically and independently far in the past, one for bodies that move around and the other for bodies that think, it is possible that they refer to a common term and concept that includes both life and intelligence as one, the way elephants include legs and tails as one organism.

Every time people coin words empirically, they remain within the example with the blind men and the elephant, since the empiric itself stands at the first existential level. If you add to this the consensual, allowing the blind to agree with the blind in a fiat manner about everything that they assume in an entire world, you lose accuracy and consistency further. This is why we persist to research life from all existential perspectives and in all its forms and realities, in order to understand life accurately and intelligently, at the third intelligent human level.

We can always equip our robot with a computer, allowing it to interact with the entire world exactly as all driverless vehicles do, at the first algorithmic level, yet computers are not intelligent but algorithmic, or repetitively intelligent, since they use only first level basic algorithms *if-then* and *repeat-until*, and cannot develop further to offer intuition and or intelligence, unless we change the computer matrix altogether. Therefore, it is better if we use a different cognitive matrix from the start, and by doing so, we set in place the actual type of artificial life that we want to create.

This is still possible, because humans had already created

computers at the first algorithmic level, which is still artificial life and artificial intelligence but only of the first algorithmic mechanical repetitive level, yet this is not enough for us to understand life. Life starts with the second intuitive level, and therefore we must create a living being of the second intuitive level or higher in order to understand life.

As stated, we might even succeed to create a different type of computers, a different type of artificial life, not based on mechanical switches but on something else, something easier to manage, in order to help us understand better matrices of life, types of life, intuition, intelligence, interconnectivity, realities, existence, natural laws, supreme characteristics, and life altogether.

However, with the current science incapable to offer us an accurate model for the intuitive intelligence, we must find it ourselves and we must integrate it in this model. It is highly complex, yet intuitive thinking is part of the development of life anyway, a significant part, and therefore we must know it in order to understand life. Intuitive cognition is animal, while humans use intelligent conceptual reasoning, throughout complete intelligent mental models. Only humans are intelligent, while animals, microbes, and even plants are intuitive. However, the human midbrain is capable to offer second level intuition at a level superior to all animals, making humans both intuitive and intelligent simultaneously. Furthermore, humans also have a first level reflexive brain at the base, the basal ganglia, offering very fast reflexes that can always save your life. This is why humans have three brains one on top of another on top of another, in order to assure a simultaneous reflexive, intuitive, and intelligent cognition.

The difference between the second level intuitive thinking and the third level intelligent reasoning is that humans perform their reasoning by using intelligent concepts, alongside intuitive images, ideas, and feelings throughout reasoning. While animals use only basic memories coupled with feelings, but no intelligent concepts as abstract as these can be, because animals do not have a cortex capable to allow intelligent conceptual reasoning, while humans can use both intuitive and intelligent thinking even simultaneously, through two different brains.

As a reference, you must use your intelligent reasoning fully while reading this book, because all concepts presented in this book are at the third intelligent level, and you cannot master them through your intuitive midbrain. However, you cannot read romance without using your midbrain, because all your feelings are in your midbrain. You can always analyze and interpret all characters and the entire plot in a rigorous intelligent manner through your cortex, yet this is not exactly the purpose of the entire romantic novel. You cannot read romance only with your cortex intelligently, but only with your midbrain intuitively. It seems that it is easier to offer our created living being third level intelligence than second level intuition, because we must integrate feelings in a second level intuitive cognition, which might be more tedious.

In order to understand thinking at all levels and therefore to understand life, we must understand both the environment and the specific intelligence coping with the environment. We live in a universe governed by random events, as random as they can be in this world. Simultaneously, entropy, which relates to disorder, increases naturally, and this is why your car becomes dirtier and your room becomes messier in time, because entropy or disorder tend to increase naturally in the universe, with Life herself decreasing the entropy continuously while keeping everything in equilibrium, by coping continuously with this entire world.

More precisely, as long as you fail fulfilling your natural living needs and meanings, by fulfilling your addicted and consensual needs instead, you increase entropy in the universe. While every time you fulfill your natural intelligent human needs in an intelligent manner, through Life and for Life, you decrease entropy, to a point when you can maintain a continuous equilibrium, when you have everything under control.

If you ever reach a point in life when you cannot handle anything anymore, it means that the entire entropy around you is too high for you to manage. While if you do not know why, just start with your addictions, by removing these from your lifestyle first, because in this manner, you can advance to the third intelligent level, when you can manage better everything, while reducing entropy considerably. While if it is not you actually addicted, but someone closer to you, these problems are so common in the current consensual addicted world, that they already define human life better than the current consensual science defines it as people moving around.

This is why entropy does not increase infinitely, because there is life in the universe, acting the opposite way, finding solutions to all random events thrown in by the randomness, the harmful, the meaningless, or the irrelevant in the universe, decreasing the entropy continuously and consistently, in an intelligent manner.

This is simplistic reasoning, because the environment itself is alive, and therefore we always have life helping life and life coping with life throughout this world and in the entire wider world, as it happens in society with people and with groups of people helping each other or competing with each other. Therefore, at the first consensual level, whatever one side of life considers order and subsistence, the other side of life interprets as disorder, environmental problems, and therefore entropy. Life against life, while this is always the case in an undeveloped consensual world.

Life takes place at the confluence between realities, which means that, in order to have life, you must have a physical body capable to hold intelligences within its inner realities. By switching perspectives from one reality to another, you distinguish either life or intelligence. Life in its basic form consists of a physical body moving around in the current reality, while the intelligence within does all the thinking as it is held by the physical body in its inner reality. Both life and intelligence match existence throughout its existential natures: subjective, objective, and highjective. Therefore, when you switch perspectives to lower realities, you find intelligences being alive in themselves, since they manage to open more inner realities in order to be able to reason.

How does existence define and affect life? Everything and everyone exists, since if they did not exist, we never had them in this world, and we never knew about them, since they never existed. More precisely, the term existence is capable only to point to, to distinguish from, or to define trivially what exists from what does not exist, in reference with an origin. Existence is only Boolean in nature, having only two states, the existent and the nonexistent. To be or not to be. Additionally, existence is relative, with one existence associated directly with each system of coordinates, as these are relative to each other. You might assume that it is similar with life as it distinguishes between the living and the nonliving, yet life is considerably more complex than existence, while we do not confuse here existence with reality. Therefore, existence defines everything as existent or nonexistent in rapport to any system of coordinates, since existence is relative. While as you study closely the natural law of existence from the base of this world, you notice how you cannot have the existent and the nonexistent simultaneously, but only one or the other, making everything that exists unique and therefore accurate.

This is how existence defines beings, living and nonliving, with each one having its own existence. This is the case because, even if some of those beings might not exist for us, they can always exist for themselves and for those around them, and therefore they have their own existence defining them wherever they are. In general, realities define everything that exists objectively within them. Therefore, if other beings do not exist objectively from our perspective, they do not exist in our world. If it happens that they are not even part of Life, this means that they are not part of our wider world in all its realities either, and therefore existence cannot define them. I define Existence to be the existence defining Life. You are part of Life, along with this entire reality, and therefore Existence defines you. Additionally, you have your own existence defining you, which is still part of Existence, since existence is relative.

As stated, existence has three relative natures: subjective, objective, and highjective. Everything within this world exists objectively, while everything within the inner realities of this world exists subjectively. This means that all intelligences living throughout the multitude of minds are subjective and cognitive in nature. They still exist, but only subjectively. Since Existence defines Life, then all intelligences are alive and part of Life, but only subjectively from our perspective. While from the supreme perspective of Life, all her living beings exist subjectively, including you, the human society, and this entire world. From her supreme perspective, all living beings of Life are her intelligences. It is similar for our higher realities, since from our perspective, everything exists highjectively up there, including all higher beings, higher selves, souls, and all religious and spiritual higher beings, while from their own higher perspective, all living human beings are intelligences.

Why exactly does it take at least an intuitive intelligence to cope with the environment? Because only intuition is original, capable to generate an infinite number of unique solutions to cope with a perfectly random environment or universe, and not repetitiveness, which is limited in solutions, and easily overtaken by the environment. Yet in practice, we notice how even capable brains fail to offer successful solutions through intuition, which is the second animal level, and even through analytic, intelligent reasoning, which is the third intelligent human level. It takes higher levels of thinking and abilities to cope with the environment sometimes, and this is the case because the environment itself comes on various levels of difficulties, many times higher, from zero to the tenth level for the supreme environment or the wider world.

As a reference, humans should be able to cope with third level environments, matching this world in order to fulfill their needs. Yet rats, roaches, and ants will be around longer than humans, since thinking alone cannot assure survival, but only life can. While throughout even more difficult environments, you need higher classes of life and higher levels of life to survive, subsist, and develop.

Life

If other environments do not allow organic life, then life has to exist there in lower forms of life, as ionic, molecular, and cellular forms of life. Microbes will be around after all ants, spiders, and roaches are extinct. The organic form of life has at its base the cellular form of life, which has at its base the molecular form of life, which has at its base the ionic or plasmatic forms of life. All intelligences are based on ionic or plasmatic forms of life directly, and not on organic life, which is only a higher level structure of life, built on top of the plasmatic, ionic, molecular, and cellular forms of life. Cellular life starts with cells, with cellular membranes. Life existed long before cells on Earth, as communities of ions and molecules as proteins, enzymes, and later on RNAs. All these are very developed ionic and molecular forms of life, with all viruses included.

As a remark, if our world is not random or slightly not random, this is a sign that it is not naturally made but artificially made, which is certainly limited not only in space, time, and field, but in randomness, existential resolution, and therefore in variety of events. Notice that even a very developed repetitive intelligence can be considered intuitive in a created, nonrandom reality, because its set or preloaded solutions, if it is large enough, it can match all events provided by the artificial environment, since they are not entirely random.

Forms of life are gatherings, unions, or communions of living beings living together in a specialized manner. Ions form the ionic or plasmatic form of life, as you find them in stars, fires, ionosphere, and everywhere else even at a lower temperature, even in space, and everywhere on Earth, standing at the base of organic life. Because ions can gather into molecules by the dozens, hundreds, thousands, or more, to exist or live life together in the electromagnetic field, while forming even larger molecules, as proteins, RNA, enzymes, cellular membranes, tubules, and filaments in a new form of life, the molecular form of life. This is the second form of life, the molecular form life. These ions and charged molecules gather as cellular components by the zillions to live life as an entire cell, and I refer to this as the third form of life, which is the cellular form of life. At their turn, some prokaryotic cells along with a zillion cellular components gather to live life as an eukaryotic cell, and this is the fourth form of life, or the eukaryotic form of life. At their turn, trillions of eukaryotic cells gather in communions to live life as an organism, and this is the fifth form of life, which is the actual organic form of life. Humans form societies, civilizations, humankind, and entire ecosystems, which are higher classes of life when left to interconnect freely and naturally, according to all natural needs and feelings. This should be the sixth social class of life, yet this is not the case currently with humanity, since the Consensual Matrix does not allow the social class of life or the intelligent human society, which should have been the sixth class of life on Earth, now dead, consensually dead.

Can you live life alone? No. As far and as much as humans can perceive and understand, life is lived only in entire types of life, forms of life, and classes of life. You can never find individual living beings living life alone, but only in types of life, forms of life, and classes of life, where they are always specialized while fulfilling their own specialized meaning throughout life. If they ever stop fulfilling their needs, then the entire form of life suffers. Therefore, whenever you stop fulfilling life and the world, the entire world suffers and even dies, as it happens in the current consensual world, regardless of what you are told.

This is how life goes through a sustained, progressive development even by having to fold upon itself throughout newer and higher forms of life, in order to become more successful. Yet this is only the physical perspective of the physical body, while this is only the physical perspective of the physical interconnectivity of all physical bodies, which is mostly empirical in nature, while we must study life through all its perspectives, not only empirically.

Moving to a cognitive perspective now, which is the inner, subjective perspective offered by existence, life creates or it gives birth to newer and higher forms of life only to allow the multitude of its intelligences to interconnect and form on their own entire cognitive systems, which are the common minds. Because while living beings make possible forms of life at their own objective level, their intelligences follow closely and are capable to form higher systems of intelligences in this manner, which are cognitive systems or minds.

There is a very precise correlation between forms of life and cognitive systems, since it is not only a matter of physical bodies holding intelligences within as cups hold water, because as stated, while from an objective perspective you have bodies and entire forms of life, from a subjective cognitive perspective you have intelligences, cognitive systems, or minds. Therefore, bodies do not hold intelligences, but bodies are these intelligences entirely at an inner, subjective level, while intelligences are the actual body from the outer, objective perspective. Intelligences are also specialized within cognitive systems, while you can never find individual intelligences that are not a cognitive system in itself.

Life has to fold upon itself into higher forms of life and classes of life only to have these. Everything relates with the specific cognitive systems that intelligences form throughout higher and higher forms of life, since their bodies and forms of life are only their outside, objective appearance. Intelligences live by the zillions within minds or cognitive systems, where they manage to ascend from the trivial, first level intelligence, awareness, and thinking, which is the basic algorithmic thinking using *if – then* and *repeat – until* algorithms among others, to the second and third levels of thinking, which are the intuitive thinking found in animals, and the intelligent conceptual reasoning found in intelligent human beings.

Humans are always capable of all lower level cognitive abilities and activities, as the first level algorithmic thinking called logic and the second level thinking called intuition, because humans have three brains one on top of another. The first brain is reflexive, allowing the basic first level algorithmic thinking. The second brain on top of the first one is intuitive, allowing second level intuitive thinking through memories that are always coupled with feelings. While the third brain is the human cortex, standing on top of the first two brains, allowing the third level intelligent conceptual reasoning, by being capable to form, hold, and maintain all third level intelligent conceptions necessary throughout the third level intelligent reasoning taking place mostly through intelligent mental models. This entire mental model of life that we are conceiving throughout this book is a very complex third level intelligent conception, conceived right now in your own conscious mind spanning the cortex. You do not exactly build it, carve it, or form it in your cortex, but you conceive it or you give birth to it, in your third level intelligent conscious mind.

As you notice, we cannot form a tenth level mental model of life as it should have been the case, since life itself is a tenth level concept, because we lack the tenth level cognitive matrix necessary to form, hold, and maintain tenth level supreme conceptions. Therefore, we have to manage to understand at least the third level intelligent life throughout this mental model of life from this book. Which means that we should be able to understand well the third level human life, while still having a continuous glimpse of the tenth level supreme life along the way.

You have the algorithmic cognitive ability of the first level, intuitive cognitive ability of the second level, intelligent reasoning of the third level, and furthermore, if you can ever go further with your development and abilities. These are possible throughout more developed brains and minds, capable to hold zillions of inner realities filled with zillions of intelligences caught in the same tedious cognitive activities as they are seen from our objective perspective, while these intelligences are caught in their normal, casual life throughout their own inner mind worlds, as seen from their own perspective. This is how everything manifests into needs, thoughts, feelings, reasoning, ideas, and intuition, as they are seen here in the real world.

Natural living human societies are capable to interconnect

freely through the natural needs and meanings of all human beings, offering all higher physical, cognitive, or spiritual abilities to everybody. This adds human meaning, strength, and fulfilment to your own existence, ascending it entirely in development to the fourth superhuman level. Yet without a natural intelligent human interconnectivity, and therefore without a natural, living, intelligent human environment, you have to take drugs in order to feel fulfilled, and it still does not work. Everybody takes drugs and it never makes anyone feel better on a longer term but worse, while this is a zero level existence, entirely outside of Life but in the Consensual Matrix, where it is considered a first level consensual existence as a brand, trademark, or consensual corporation.

Higher forms of life and therefore higher minds or cognitive systems can offer physical and cognitive abilities, or not, depending on circumstances. Cells can always outlast entire organisms and entire species throughout the most drastic calamities, even though higher forms of life and higher cognitive systems are significantly more capable than cells. This is the case because cataclysms and unfavorable environmental conditions in general affect niches in various manners, while all forms of life occupy different, distinct niches. Cells are of a different form of life than organisms and entire species, being affected by the unfavorable environmental conditions in a different manner. While cells themselves cannot survive freezing temperatures but ions can, since the cellular form of life lives mostly in liquid environments, while ions can live everywhere in the field, even out in space. The more unfavorable the environment is, the higher is in level, and the more capable you have to be as a living being in order to withstand it, since you have to match it with your own level of abilities. Yet with an entire humanity kept undeveloped, everybody fails.

We have also discovered a more natural meaning of life, as to cope with the environment, to interconnect naturally, meaningfully, and harmoniously within a natural, living human society, and to learn and develop in every possible manner in order to be able to cope with the environment, which includes society and the human mind, and not only the physical world.

Therefore, we can decide for our robot its meaning in life. We decide to allow it to survive, subsist, and develop naturally, and to do so in any manner it chooses. Therefore, its meaning in life is to cope with its environment, and ideally, to learn more and to develop in order to cope better with its environment. However, just as leaving wallets unattended, eventually, someone can find our robot and temper with its meaning in life, making our little robot work all day long on behalf of others, as it happens with humans.

You do not actually have an essence of life, even though you can always call it so, but you have an entire matrix of life spanning the entire world. Therefore, if you ever want to create, find, or become alive, if you ever want to transcend from the nonliving to the living, there is not exactly a substance, a cell, or a bone that you must add to your objective material physical body in order to make it alive, but you must be already in the matrix of life, you must be already alive. There is no essence of life hidden somewhere, because life entirely is the source of life, the essence of life, the matrix of life, and life itself.

We still attempt to provide our robots with the living intelligences of life to use as a mind in order to be able to survive, subsist, prosper, and even develop on their own. It is still possible that our model of a created natural living being will prove us wrong, and we will manage to find the magic essence of life, or to create these natural living intelligences from scratch, allowing our robot to become alive.

A solution would be to attempt to build an entire natural intelligence from scratch to equip our robot, which is amusing, since we end up building one living being in order to build another. Why don't we just build a natural intelligence in a computer program, as in a videogame, and forget about the robot? Because we end up building a subjective living being in a computer program, while we want to build an objective natural living being, part of this world, in order to see it better.

Life

We want to see, feel, and perceive life, the matrix of life or the essence of life as an ingredient here in this world, while this can be done only here, in our world, but not within computer virtual worlds, where everything seems ambiguous and can be understood and interpreted in any manner.

Regretfully, we have to discard the computer, since without the created natural intelligence, our computer is just a common artificial algorithmic intelligence of the first level, an automatic processor guiding an automatic machine, which will never create a natural living being, but just another machine.

We are back to where we started, but now we know what we need in order to create a living being. We have to make it capable to react to every change in its environment, known and unknown. Life has a multitude of meanings besides survival and subsistence. Living beings have developmental needs and meanings, along with interconnectivity needs and meanings, and much more. We are only keeping our model simple by assigning to our robot a subsistence meaning in life and in the world. You will see how complex our model has to become, only to be able to fulfill this meaning of survival and subsistence in life and in the environment as a created natural living being.

Our failure so far is typical to any mental model, because this is how mental models are always made throughout intelligent reasoning. You fail throughout your reasoning, then you identify your reasoning problem, you try again through different methods, mental model after mental model, you learn with each step of the model, while you restructure your reasoning to make it simpler and easier to follow. By the time you obtain your successful idea, you fail and you redo your mental model a multitude of times, yet you know exactly what to do in order to solve your problems while understanding them entirely.

Intelligent mental models are part of the human comprehensive intelligent reasoning. Mental models are third level intelligent reasoning, while in order to perform them, you need a cortex, along with a diverse abstract conceptual language and accurate understanding of everything that you need to know, along with a detailed, accurate intelligent inner replica of the world made of a multitude of accurate intelligent conceptual memories about the outside world involving your subject of study, called intelligent conceptions.

Therefore, in order to obtain third level intelligent reasoning, you need a third level intelligent society, filled with a multitude of developed robots, while we are still at the first level of algorithmic, repetitive thinking, with only one little robot that still breaks down constantly. Yet we are getting there, and the more we stumble into problems, the more we learn and persevere, since this is the case with all intelligent mental models.

We notice something important in our model of a created natural living being. First, at the beginning of the model, we built only an animated physical object capable to jump around in circles and therefore move around, and it was not alive. Then we assumed that intelligence is the essence of life, we searched for it, and we could not find it outside the matrix of life, concluding that you always have to be alive in order to become alive, in a trivial manner. Yet now, as we persist to create a natural intelligence as an actual essence of life, we realize that meaningful interconnectivity is actually at the base of natural intelligence, being the essence of natural intelligence and furthermore the essence of life. While we find the same interconnectivity taking place within cognitive systems as they take place in society at the outer objective level of the physical body. Furthermore, it is the same interconnectivity transcending realities with each task and fulfillment, and this is always the case in life.

Intelligence, cognitive systems, and minds in general are subjective, inner, and cognitive, compared to the objective physical bodies holding them within. Furthermore, interconnectivity, as the basic interconnectivity taking place among all intelligences of any mind or cognitive system throughout reasoning, this cognitive interconnectivity has an inner, subjective existence compared to the cognitive system itself.

We are going lower and lower within existential levels while attempting to find life, the living matrix, or only the essence of life, whatever we can find. Yet with the entire world formed of countless of inner realities many times one within another within another within another, we might never have the chance to make it all the way down to the source of life, essence of life, or matrix of life. Yet once we find the elemental intelligences, the specific intelligences that are not cognitive systems in themselves but distinct individuals, we reach the last inner reality, and we are there at last, at the source of life.

From what we know so far in the model of life, there is the living Field all the way down, holding life entirely from there, capable to hold the entire matrix of life within its continuous, living, intelligent, interconnective encoded vibration. Can anyone ever tap into that in order to become alive? Will this model of a created natural living being, along with our entire model of life will ever take us there? Yes, but only if we reason accurately and intelligently.

What exactly is an intelligent mental model? Can you not have a definition instead, or an organized enumeration of statements and concepts helping you understand life? No, because all lists, enumerations, graphs, theories, and assumptions are only at the first level of knowledge, using logic, basic inductive and deductive algorithms or even less, using only simple beliefs, theories, and stereotypes forcing you to believe whatever they claim, while these are not enough to sustain an entire third level intelligent rationality. While life itself is associated to a highly complex tenth level living conception, and you can barely reach to understand it even from your third intelligent human level.

At the third intelligent level of cognition, you use entire cognitive or mental models, which on their turn use entire comprehensive intelligent cognitive interconnectivities taking place among zillions of inner intelligences spanning the cortex, as they mimic or model accurately all world details and circumstances, helping you perceive, understand, elaborate, and memorize everything. If you ever experience persistent, powerful thoughts going on in your head on their own while keeping you awake at night, this is them, your concepts and conceptions spanning your cortex and middle brain, reasoning continuously for you, while keeping you awake just as well as a conscious intelligence, to reason alongside them, in order to solve any problem.

It is never enough to present a definition of life or an entire table of statements characterizing life in order to explain life, because life itself is of the tenth supreme level, understood by humans at their own third intelligent level. This is still sufficient, since you end up understanding the intelligent life around here, mostly the third level intelligent human life, yet you have to explain and understand everything through third level intelligent conceptions, part of third level intelligent mental models, which never fit in definitions and tables of statements about life.

In contrast, when you persist to think through beliefs and stereotypes at the first stereotypical level, you remain entirely at the first cognitive and behavioral levels, below the intelligent human level. In this manner, you miss your intelligent human fulfillment and meaning throughout life, as Life does not even consider you part of the matrix of life, but nonliving and even nonexistent, since you fail to connect with the matrix of life through your cognition, decisions, behavior, and inner and outer interconnectivity.

At this stage of our mental model of a created natural living being, the essence of life relates mostly to the intelligence required to render our robot capable to respond to every change in its environment, everything, including the little rocks causing it to lose balance and fall, rust and sand accumulating in its hinges blocking its movement, and dust and dirt covering the solar panel limiting its energy. Our robot needs intuitive learning to form its intelligence and cope with everything, and therefore we must equip it with a created natural intelligence, again. Which is extremely difficult now without the computer, since we have to develop a created natural intuitive intelligence or a rational intelligence from scratch. Which is very good, because by doing so, we understand intuitive thinking and rational intelligence just as well.

As a reference, logical, algorithmic, and ideological thinking are of the first cognitive level, and can be reached by natural and artificial digital intelligences, and even by basic calculators. Intuitive or animal thinking is of the second cognitive level, and you have to be alive in order to achieve it. Furthermore, comprehensive conceptual intelligent reasoning is of the third intelligent level, characteristic to living human beings, through their powerful cortex on one side, through their higher level intelligent social interconnectivity on another, and through their higher level objective and abstract intelligent human knowledge just as well.

We can build an abacus, which is a mechanical objective computer, using specific sequences of holes made in cardboard punch cards, for every change in the environment that we can imagine. Yet we use a cardboard punch belt instead, as they use in mechanic pianos or in expensive automatic antique toys. These are also simple computers, capable to hold only basic artificial intelligences, while in order to make our robot adapt to any circumstance, we have to invent a natural, living intelligence, present here in this world, which we cannot.

As stated above, life and intelligence are more complex, part of the same concept, with life seen at the level of this outside world, and with intelligence or consciousness present within the inner mind worlds. Furthermore, once we drop another existential level below intelligence, we find interconnectivity, which is already part of life and intelligence. More precisely, throughout our research of life, we find intelligence when we advance in details, while throughout our study of intelligence we find interconnectivity while advancing in details. Which could mean that the essence of life is intelligence, while the essence of intelligence is interconnectivity, and then the essence of interconnectivity is life again, the matrix of life.

Furthermore, through interconnectivity and specialization,

we can create higher forms of life, and these define life, but at a higher level or higher form, in a living existential spiral that should define one single concept but there is no word for it, and we have to define it through life, intelligence, and interconnectivity, as one. I refer to it as oneness at any level, and I refer to it as the One at the ultimate, absolute, supreme level of Life.

Can we do the same with a created natural living being? No. You have to be already part of the matrix of life in order to be alive, and once you are part of the matrix of life, you are always alive, with death itself only switching you to lower levels and lower forms of life, all the way down to the electromagnetic field itself, if this ever becomes the case.

For this world, the field holding the matrix of life in this world is the basic electromagnetic field that you already know, making possible the electromagnetic type of life. This might or might not be the case for the entire wider world, since there can be a multitude of types of life besides the electromagnetic type of life.

Through the matrix of this entire world, now Life, Intelligence, and Interconnectivity are capable to transcend to all lower and higher realities everywhere, as they already span the wider world, and as they are the wider world entirely, depending on levels of perspectives.

We have been here before, losing life into nonliving matter, and now losing life into intelligence and interconnectivity. We learn in school that in time, nonliving matter creates life through evolution or even through spontaneous creation, and then after some more time, life creates intelligence, or this is what they teach in school. Therefore, life should dissipate only into the universe, while intelligence should dissipate only into life. Yet we always find out that everything dissipates into everything when we change perspectives from one into another, or when we assume supreme perspectives at the largest and smallest scales.

What is life? What is intelligence? What is the universe? How exactly do they diffuse, transform, superimpose or transcend one into another? More importantly, what exactly is that 'elephant' that they form together? Because we are now exactly in the circumstance of the three blind men perceiving an elephant from multiple perspectives. We are blind because we are restrained to our world in perception and understanding, as we are trying to understand several supreme perspectives at once: life, intelligence, interconnectivity, causality, the field, existence, and the wider world, from here, form this world.

If we have to focus not only on life and intelligence, but on interconnectivity and physical bodies instead, then let us build more robots. Yes, let them be identical or specialized, and let them be able to interact and help each other get back on their feet again when they fall down, since this is how life takes place in the real world at all existential levels. Everything is interconnected, and many times, everything is harmonious.

Let our created living robots be able to clean each other's solar panels, lubricate each other's hinges, and furthermore, let them share solutions to all encountered problems, just by being able to download relevant information from each other's punch cards and punch belts. We have the freedom to build as many robots as we want, either similar or specialized in various tasks: the cleaning robots, the lubricating robots, the robot repairers, the robot part builders, the robot assemblers, and the robot researchers, all holding the blueprints of their mechanical tasks in their punch cards or cardboard conveyer punch belts.

This is how we build the originals, the prototypes, and then they can create their own parts and replicate themselves as they wish. They can do so by using solar energy and raw materials, whatever is already available everywhere, before they learn and are able to gather raw resources. Furthermore, let there be specialized robots able to copy and inscribe codes on new cardboard belts from scratch, or just repair damaged and worthless sequences of holes for each robot. In this manner, robots use only successful coding of data for overcoming problems, discarding worthless information, while sharing conclusive, verified feasible information among themselves, now called experience.

We note how even by processing information not through a regular computer but through a mechanical computer using punch cards and punch belts, the coding itself still forms an inner reality, since the processing itself using this coding exists subjectively, in an inner reality. We still have the physical objective bodies and the intelligences guiding them, and these form life. We are still at the first level of thinking, the simple algorithmic thinking, yet through this new setting, our robots have the chance to survive and develop, in order to learn and develop themselves to the second level of life, the intuitive life.

Does this model of a living being or living community really work? Have we built a living community of robots? Yes we did, and it has survived for days now. However, there are some remarks that we have to make, regarding their intelligence. Our robots still do not have an intuitive intelligence, and therefore their days are numbered. As stated, intuitive intelligences are capable to learn, update, predict, and adapt. Intuitive intelligences are capable to generate an infinite number of responses necessary to cope with an infinite number of random events offered in time by the environment.

Our robots are alive by very low standards of living, and they are still at the first developmental level. Even at this first mechanical level, they have a very low status, as they still exist only by chance. Yet this is also the case with humans, just study history to see it yourself. By sharing information and solutions to cope with the environmental events and social events, our robots manage to increase their chance of survival and subsistence in their environment. Yet in theory, they are still not capable to cope with all changes in the environment, since their number of responses is still limited, whatever we have managed to preprogram, plus whatever they had learned and experienced so far. They can further increase their chances of survival while still using their repetitive intelligence, by drastically increasing their population, and therefore by drastically increasing their successful environmental solutions that they share among themselves.

Can robots evolve? The theory of evolution has only one statement, species evolve, while it also gives a classification of species on Earth, in order to prove itself right. A later theory, the survival of the fittest, explains that species evolve because the unsuccessful species die away, and therefore the survival ones remain alive, which is what this theory states. Yet it does not explain why the survival species remain alive, nor how these happen to evolve, because if it did, now it actually helped us in our study. Therefore, while the theory of evolution has one statement, species evolve, the survival of the fittest theory also has one statement, unsuccessful species die. These are both trivial, and therefore useless even for our model of a created living being. However, while modeling life, we have to figure out what is wrong with these, only to be able to give the right explanations throughout our model.

Why do species evolve? The answer is embedded in the meaning of life. Our general answer so far is: because Life wants it in this manner, yet this is always the case from absolute, comprehensive perspectives. In a following chapter, we will see how Life, Intelligence, the wider world, Interconnectivity, the field, and the Divine are distinct perspectives of the Supreme Living Being, using them according to your domain of study or beliefs. Therefore, at this comprehensive perspective covering everything that exists, the answer to our question about the meaning of life is: because the Supreme Living Being wants it in this manner.

This is the unending dilemma and debate between creationism and evolution, not a matter of what is right or wrong in these beliefs, but it is only a matter of perspectives. If creationism is true from a comprehensive perspective at the level of the Deity itself, evolution tried to model the development of organic life on Earth during a few billion years here in this world, with only one statement, that species evolve, which is not even an answer. It is only a difference in perspectives, and therefore creationism and evolution are never compatible to be able to compare themselves even throughout first level debates. Since this is why this debate goes on, yet only for as long as people remain underdeveloped, at the first consensual ideological level matching them.

There is a distinct status associated with each being, community, nation, culture, civilization, and planet. You can have a mechanical status, a servitude status, a consensual status, an animal status, or an intelligent human status. Your status is different as a mechanical, consensual, ideological being of the first level, than as an intelligent living being of the third level, since you have significantly less abilities and rights at the first level. Yet you do not have rights at all at the first level, since at the first level, you are not considered alive on your own, but only through your masters, owners, or profiteers, and through their status and life. This is why you have no human rights as a corporation, brand, trademark, slave, soldier, employee, or servant, but only privileges, whatever they give you. You have no human rights when you are your consensual corporation, since it is consensual, and not a living being. It is the same for communities and for the entire Brotherhood, society, and civilization, since these are jurisdictions, ideologies, and corporations, direct part of the Consensual Matrix, always at the first nonliving level.

With the level of the current science and therefore current human knowledge associated to the consensual status of the human civilization, humans admit explicitly and implicitly that they are not exactly an intelligent reasoning civilization, not even a living civilization, but more likely a mechanical or consensual civilization of the first level. Therefore, the human interaction with natural, intelligent civilizations can be limited, restricted, or restrained, due to inferiority or incompatibility, placed at the bottom of all life chains of the wider world, and always within the Consensual Matrix. Rights start with the second level, because rights start with life, while privileges are only at the zero and first levels, given to nonliving beings.

Why having so many statuses in the wider world? Probably because life is diverse: living, nonliving, organic, spiritual, virtual, mechanical, artificial, unconscious, conscious, universal, microscopic, plasmatic, ionic, crystalline, and stellar, all categorized on various levels and degrees of statuses, rights, and privileges, according to the living being's abilities, age, origin, intelligence, knowledge, awareness, organization, form, thinking, development, lifestyle, beliefs, choices, society, laws, structure, meaning, and behavior, determining their unique status, from zero to ten. While you need a very diverse set of statuses and laws to manage everything in an egalitarian, harmonious manner.

No being can own, harm, use, exploit, employ, or order another being of higher or equal status without its explicit permission. You are free to do so only with an inferior being if you need, or if it fits your meaning, yet even then, you cannot harm unnecessarily the inferior being in any way. This is also the case with any type of community and civilization, and even with any interaction between individual beings and communities of beings, since there are not many differences between individuals and communities of individuals when communities of individuals behave as individuals, as an organism, or as a harmonious civilization.

Yet the law given above is not a higher law, but only its interpretation. The higher laws do not give you your status and rights, but they only respect them, since you have your status and rights from birth through your abilities, achievements, and development in life. The higher laws only make order in all chains of life through a genuine hierarchy of life, from lower level status to higher level ones, to become the order in all chains of life, including food chains, informational chains, social chains, developmental chains, and habitat chains. It is a law of discrimination of life in the wider world, with inferior beings of lesser status on the bottom, and with superior, privileged beings of higher status on top.

This is why currently, you can eat chicken but chickens do not eat humans, while both species are omnivores, and therefore both species can eat each other. Because through the higher laws, you are superior to chickens, you will always maintain yourself above them in your common food chain and existential chain through your development, knowledge, awareness, and abilities, yet if the chickens ever manage to find ways to climb above you on the common food chain, then powerful higher intelligences can intervene to restore order. They do so only for as long as the chickens remain one status level below you, at the second animal level, with you at the third intelligent human level. Because if you are ever equal in status, if you are both animals as science and society state, then no one will ever intervene on your behalf.

Yet if you identify yourself with your consensual corporation instead, as your name written in uppercase letters from your ID card, then again, no one will ever intervene, and the specific chicken, animal, consensual, or mechanical intelligence claiming to own you can do with you as it pleases, for as long as it fulfills its needs and meanings in the world. You are even inferior to chickens as a consensual corporation, and this is why it is not even considered that you eat chickens as a consensual corporation, you are not even allowed to eat as a consensual corporation, and you have to change jurisdictions only to be able to eat and visit the bathroom, in private, in the natural, real, living world. You must respect the court and you can never eat in court, because you are only a corporation in court, and if you ever eat in court, you become a living human being, and the entire court dissipates, because it cannot hold living human beings in its consensual matrix. This is what the tyrants invented thousands of years ago for exploitive reasons, and it still works, while deciding the faith of everyone in this world.

How can you, by holding a lower status, exploit a more developed civilization, and still get away with it, by any higher law? Simply, you lower the status of that civilization below yours, in every way. You claim a lower, unconscious status for that civilization and claim it as your property, since you found it or salvaged it. If not possible, you can still convince, fool, trick, or force that civilization in every way to agree to your intentions and drop into servitude, by claiming to be consensual, only a corporation, always at the first level, as it happens everywhere throughout the Consensual Matrix. There are many ways to twist the higher laws, because the higher laws apply only to living beings, and only if there are no other agreements in place. The higher laws respect all agreements that you ever make, because the higher laws respect your entire status, decisions, agreements, interconnectivity, life, reasoning, needs, purposes, fulfillment, lifestyle, behavior, and intelligence.

For example, declarations of war are simple agreements, voiding all agreements made during peaceful times, because the entire mode of society shifts to the social mode of war. Therefore, all laws used in society and in all jurisdictions change to their martial state, using martial laws, while this allows humans to kill themselves during wars without prosecution, this is their decision and mode of interconnectivity, and the higher laws along with all higher beings up there respect it unconditionally. Currently, many powerful higher entities cannot kill humanity, since humanity is not in their existential chain, just the way you cannot kill and eat golden eagles, but only chickens and fish, since these happen to be in your food chain. The higher laws state clearly that you have the right to fulfill all your natural needs and meanings for as long as Existence defines you throughout your specific natural environment.

These powerful higher entities can always determine humans to shift to war modes of living, fight and kill each other, then salvage humanity entirely, and this is another way to twist and therefore evade higher laws, along with a multitude of other strategies. This happens repeatedly, with the human civilization and even with nations and individual humans, with you an example every time you associate yourself with your corporation.

These higher laws apply to all living beings, and not only to powerful entities or civilizations. Coincidentally, these higher laws are more implemented in society. This is why you cannot get a job if you do not identify yourself as a corporation, which is a nonliving entity of the first level status, and you do so by registering it first, creating your own corporation in this manner, with all the necessary demands: proof of consensual identity, registration number, explicit request, and signature. This happens when you apply for your social insurance card or social security card, since you always need one to work.

Did you ever wonder why these legal procedures are so awkward, extremely complicated, and extremely precise, governed by laws which seem different than the ones you know in society, including specific details never concerning you and society? Because they never concern you, addressing only your owners and the higher laws, as they are meant to evade higher laws. Did you ever wonder why suddenly, you seem to have rights, you seem to count in the world, and you are conveniently protected by invisible upper laws that seem to come from outside society? While these invisible laws protect your pets and the wild animals just as well, along with racehorses and farm animals, forbidding anyone to harass them unnecessarily. Because these higher laws span the wider world, as they always protect living beings, regardless of the reality in which they exist throughout the wider world.

What exactly can our robots do to achieve their second level intuitive thinking or more? Learn and develop. Yet how can they do so, when our robots do not exactly have natural needs and feelings as humans do to make them develop, because our robots are not actually alive? We must equip them with similar needs and meanings, yet we do not even know what major cataclysms come and when they do so, therefore how can we ever program and prepare our robots, when we do not know it ourselves? Will they ever become alive? Will they ever develop these needs and meanings themselves to render them capable to survive, just by sharing successful information that they can already memorize themselves within their punch belts?

It is always the same answer: the robots should develop continuously, on their own, just in case. We wait for our robots to do so, to reach the animal level and therefore intuitive thinking, becoming in this manner naturally alive, while ending our model of a created living being successfully, so we can move to the next chapter.

Humans already have their natural needs within the human cognition, meant to develop them and the entire world to the intelligent human level. Humans should be capable to achieve their intelligent human level in one lifetime, or only throughout childhood, if they grow up in the intelligent human environment, or if they can avoid the drugs and entertainment of the zero level, the empty consensual servitude of the first developmental level, and the low animal feelings of the second animal level. Only with all the necessary accurate knowledge and only after overcoming all lower level unfavorable conditions, humans can reach the intelligent human level, if they manage to develop, since it is not an easy task. It already happens currently, within very successful families, yet the current consensual society is meant to keep humans on lower developmental levels, in every manner and through every scheme, as you can easily notice everywhere, since these keep you down while you keep others underdeveloped, if you are not careful.

This is why humanity will never ascend to its genuine intelligent human level, to free itself from the enslavement of its own kind and of any higher malign entity. Not exactly because humanity is too young, too incapable, too disabled, or too corrupted, since all natural developmental needs are already within all humans, manifesting daily, as the continuous boredom, restlessness, and even despair and madness, since this is the punishment that you receive for your failure to learn and develop. Consequently, you have to take drugs only to feel normal again, or this is what society teaches you, while you never feel normal again, but you decay continuously while feeling dreadfully.

As a reference, study your feelings closely every time you engage in intelligent human activities, as continuous learning, development, teaching, writing, and painting, to see how you never feel lonely and bored, but genuinely fulfilled, and this is how it should feel continuously as a human being here on Earth.

Then why does humanity remain underdeveloped, and therefore harmful? It is a major undertaking to keep humanity underdeveloped, with the entire Consensual Matrix working hard to make it possible, and with the entire Brotherhood working overtime to keep themselves and the Masses underdeveloped.

It is different with our robots, since it is the other way around, we try hard to develop them and make them alive and not dead, consensual, and underdeveloped as it happens with humanity and with all worlds and realities of the Consensual Matrix. We are still not capable to model intuitive thinking ourselves to offer it to them, probably because it is impossible to model natural needs, meanings, and feelings for an artificial robot. Needs, meanings, and feelings are not exactly achievements in life, but they are interconnective living cognitive abilities coming from the continuous interconnectivity of the multitude of inner specialized intelligences of any cognitive system, counting in zillions, harmoniously.

Within minds or cognitive systems, intelligences are specialized, and therefore they are capable to tend to all their inner tasks within cells and within the entire organism, counting in zillions. Yet what they cannot do themselves, they pass on to other specialized intelligences as needs, depending on their specialization, with many of these tasks passed on to you as a conscious intelligence, as your own common needs, and you have to fulfill them in the outside world. This is the fourth supreme characteristic of Meaning. As a conscious intelligence, your task is to manage and coordinate the entire organism as it interacts with the outside world while fulfilling its needs in the outside world, or while fulfilling your needs in the outside world, since it is the same, because they are the same needs. Yet since the outside world changes continuously, you have to reason intelligently continuously in order find the best solutions to fulfill your needs.

This is the case with all your intelligent human needs, and you fulfill them continuously throughout life. Everything that you do in life you do to fulfill your needs and meanings, natural and artificial. You fulfill your natural needs and meanings on behalf of Life, as you are constrained to fulfill your artificial needs and meanings on behalf of the Consensual Matrix, artificial needs and meanings that go against Life many times, and therefore against you, against your fulfillment, against your loved ones, and against their fulfillment, adding significantly to an entire dreadful human condition here in this world.

We can equip easily our robots with artificial needs, by adding these in the computational code of their punch belts. Yet as seen, we cannot predict everything, since their own existential condition in their environment is very diverse, and our robots perish with the first unpredicted calamity.

We can study the Consensual Matrix to learn how it manages to keep saving the multitude of its worlds and realities that it controls throughout the wider world, since it does everything through artificial consensual needs and meanings. This works only temporarily, until the entire world dies, through lack of actual fulfillment. We can study Life instead, to learn how to equip our robots with natural needs and meanings, capable to develop them sufficiently to help them survive, subsist, and develop under all circumstances. This is what all living beings do throughout the wider world, while everything is alive and very successful. Your own cells are always successful and even flawless while fulfilling all specialized tasks throughout the organism, while if anything goes wrong within the organism, it is mostly your fault as a conscious intelligence.

How does Life do everything successfully? Because once we understand it, we find the essence of life, we find exactly what differentiates the living from the nonliving. Life is ageless, while she always remembers all her successful solutions even from one species to another, always remaining successful in everything that she has already encountered and in everything that she had already fulfilled. We are always in the same situation, yet the organic form of life had already developed a cortex offering intelligent cognition, while we cannot even offer basic intuition to our robots.

It is easier to study the Consensual Matrix first. Throughout major critical circumstances, the Consensual Matrix simply orders its more developed beings that it subdues, to help all worlds and civilizations under its control whenever their higher intervention is needed. This is how larger asteroids never fall on Earth, major volcanos do not erupt excessively, intense solar flares never reach Earth, and nuclear bombs never explode. Dreadful events never happen, as it is the case on all farms and plantations, including the current human civilization, since all farmers tend to their farms well. Yet whenever humanity refuses its servitude, since it happened before, the Consensual Matrix ceases to help humanity, all cataclysms happen, and these are dreadful enough to switch Earth from one civilization to another.

Yet many times, the Consensual Matrix does not help humanity even when humanity serves well, even refusing to inform humanity of the incoming cataclysms. It informs only the Elite, or it informs and helps only a privileged family, a genetic line, or a specific nation, whoever is more profitable, since all these happened throughout the past ages of Earth.

Everything happens in this discriminatory manner because the Consensual Matrix persists to maintain humanity underdeveloped and therefore vulnerable to all critical conditions. Because throughout the ages of Earth, the Consensual Matrix is absent on Earth, and humanity develops naturally, capable to predict and avoid everything, mostly by using higher abilities. Humans are potentially capable of higher development and higher abilities, past the third intelligent level, as you find it mentioned throughout the old records.

Life is different than the Consensual Matrix. The current science teaches that living beings cope with their environment while causing them to develop continuously, throughout a major competitive interaction within their environment. This is not exactly true, because all living beings are direct part of their environment, many times harmoniously, as in a comprehensive dance that they maintain continuously throughout life within their natural, living environment, throughout a comprehensive interconnectivity. In this manner, living beings do not exactly fight for survival and subsistence as we try to make our robots do while hoping to understand life, since this is only what science teaches, survival and fight for life at all costs, always resulting in evolution. This is all that people know, and therefore this is what people expect in ignorance, to survive at all costs, and it never happens. Who exactly fails here in our mental model? Humanity, our robots, the current science, Life, the entire consensual society, or the Consensual Matrix? Life never fails, and neither should humanity in its golden ages.

Because in real life, throughout golden ages, living beings are integral part of their environment, filling up specific niches within their environment, while they only work hard to reach and maintain their niches throughout life, throughout the most proper interconnectivity taking place at all levels, conscious, subconscious, and highconscious. Therefore, if we want to create living robots, we have to integrate them naturally within an entire natural environment. This is already impossible, since the entire environment is already taken, already occupied in all its niches, and it is already alive and already filled up. This is why our robots fail continuously, because they remain the only created artificial elements of a living environment, never actually fitting in the environment, never even achieving to dance harmoniously with the environment alongside everybody else.

Our robots become rapidly the unfavorable conditions of the entire living environment, and in this manner, everyone and everything turns against them while causing them to fail, since they are no match for life, since Life has been doing its own comprehensive harmonious dance of subsistence and development since ever, at least in this world. While our robots are new and unequipped, going against the environment many times, because we fail to understand life so far. Yet we are making progress, so let us see more.

In the beginning of this study, we wondered how we could

ever perceive and understand life comprehensively with us inside, since we are always alive. Yet we have managed to find our way, since we introduced in our model of life two distinct outer perspectives: our lifeless robots, and the entire lifeless Consensual Matrix, alongside the multitude of inner, living perspectives at all levels, at subnuclear level, field level, intelligence level, nuclear, atomic, molecular, subcellular, cellular, organic, social, cognitive, interconnective, and absolute levels, as Life, Intelligence, Interconnectivity, the Divine, Consciousness, the wider world, and the Supreme Living Being.

How much has the current science helped us on the way? Not much, while it employs tens of millions of scientists with their flawed consensual knowledge many times getting in the way. Furthermore, these scientists take the place and meaning of capable people who would have made a change in the world, but are never employed by science to make a better world.

The Brotherhood works overtime to keep humanity this way, while taking humanity to its common grave proudly. This happened repeatedly throughout the past ages and civilizations of Earth, just study the old records to see it for yourself. You can still study what is left of them, since the Consensual Matrix hides and alters all records continuously.

Throughout life, you do not interconnect with your environment only consciously, since this is what only the current society teaches. In real life, you interconnect freely, at all levels: subconscious, conscious, classconscious, and highconscious, through all your intelligences, throughout the fulfillment of all your natural needs, in a continuous harmony and cooperation, which is the positive win-win interconnectivity. It is meaningful to maintain your interconnectivity positive continuously, because if you ever fail those around, they remember you, and they seek to avoid you in the future or even to disable you and to get you out of the way. While without them, you cannot fulfil your own needs, and you fail.

Life

This is why, even in the jungle, living beings seek positive interconnectivity with those around, at a conscious and subconscious level, for an assured future interconnectivity with them. While as we notice, it is not exactly the continuous fight for survival that you can barely manage throughout life and therefore this is why you are still alive, since this happens only in movies and in the theories of science. Everything is part of the current indoctrination, disconnection, and egoistic social interaction. Only your achievement is capable to maintain a harmonious, positive interconnectivity with those around throughout life, since you can always help each other, to achieve even more.

We study and describe only the Consensual Matrix along with the matrix of harmonious, positive, natural, living interconnectivity of life, while these two distinct matrices, as all matrices of the wider world, are capable to form, hold, and maintain entire realities themselves. Within these interconnective realities held by the matrices that we form ourselves throughout life and throughout servitude, live the actual intelligences sending all social, consensual, and living needs and meanings to you and to everybody else from the natural and consensual matrices that we form ourselves. These social or consensual intelligences send you the necessary needs to render you docile and in servitude within the Consensual Matrix, or to render you disposed to develop harmoniously, naturally, and lively, and then to specialize naturally exactly in the specific living tasks and living meanings necessary to avoid or counteract major dreadful critical circumstances, in the intelligent human society and on Earth in general. Do this to our robots now if you can.

We know the major living natural ingredients forming life, and they are intelligence, interconnectivity, the field, and the entire real, living wider world. Now we can place these in our robots, and they become alive. We equip our robots with specific living intelligences having their roots in the electromagnetic field, intelligences that must interconnect freely and naturally, since only in this manner, they are capable to send each other the multitude of natural needs, including the specific developmental needs rendering our robots capable to learn and develop adequately enough to withstand all incoming critical circumstances. This is how our robots remain capable to inhabit this world, which is a natural, living environment, and this is how our robots become alive. Job well done, in theory, but not in practice, yet.

Notice how we have already failed to create a living being, yet what we try to do now, is to help our artificial machines to become alive themselves, if we only know how. With science not too far behind, because science states that life evolves from the nonliving to the living, throughout a continuous fight for survival. Living victory and survival itself differentiate the capable from the incapable and therefore the living from the extinct, or this is what the current science states, trivially.

If we can manage to equip our robots with intelligences that tap into the raw electromagnetic field, we manage to render them alive. This is the essence of life, the raw, natural, living intelligent encoding of the electromagnetic field. We can do so by cloning, growing, and grafting to our robots a living brain, since any living brain already contains the necessary living intelligences that are already tapped directly in the raw intelligently encoded electromagnetic field.

Yet this grafted brain is already alive, not produced directly by our model, but it is only transposed to our model. Furthermore, we are not creating a living being, but we are simply equipping a living being with a robotic body, which is our robot. In this manner, we cannot understand the raw natural intelligences roaming the grafted brain, and therefore we cannot understand life, the essence of life. How are life and intelligence encoded in the electromagnetic field at the base of the ionic form of life? We must study this, yet this is only the electromagnetic matrix of life, making possible only the electromagnetic type of life from among all life of the wider world.

Our only option is to develop these natural intelligences directly by our mental model, ourselves. First we have to find and understand the essence and origins of life ourselves, in order to understand life entirely, and then in order to be able to apply our knowledge while building from scratch a living being, any living being. It is more tedious in this manner, but it helps us understand life.

These inner specialized intelligences, as your eating, recovery, reproductive, social, or developmental primal intelligences, force or tempt you through their needs and feelings to do all tasks for them, everything that they are not capable to do on their own in the outside world, as finding food in the outside world, interacting with the people of the outside world for various reasons, or finding safety when needed. Our robots have simple mechanical punch cards computers, which we have programmed ourselves, as best as we could, with a limited number of circumstances for them to follow and respond. While the natural, real conditions and events are unlimited in the outside world, and therefore it is only a matter of time before our robots stumble into unexpected dreadful natural circumstances, to break down irremediably. Our robots try to learn on their own and then they share the information, but how exactly can you learn on your own about major drastic events that you have never encountered? Without knowing exactly how to respond to these dreadful future events, you remain unsuccessful when they manifest, and you become extinct. This is the case because our robots still lack second level animal intuitive intelligence or third level human rational intelligent intelligence to help them through unknown unexpected dreadful circumstances. Yet life also remains unsuccessful many times throughout unexpected dreadful circumstances, while life always learns from all mistakes while always avoiding them. Our robots have their punch belt computers helping them remember everything necessary to survive, while this should be considered a living success, as slight as it is.

How can you ever prepare for a drastic major event, when these occur every ten thousand years or so, and by the time the next one happens, you already forget what the last one was?

Life remembers everything in its cellular nucleus while monitoring continuously the entire environment for everything significant that can happen again. While when it happens again, life sends needs and feelings to all living beings in order to behave in the most proper manner that remained successful in the past. This motivates all your phobias associated to heights, spiders, fast passing cars, confinement, earthquakes, dreadful people including monsters and tyrants, poverty, fire, deep water, dark places, lack of air, and tight places, because this is how many living beings and entire species and civilizations perished in the past, while life always remembers them and always sends you all the necessary needs and feelings to detect, predict, and avoid them. You do not have to take medication in order to overcome your phobias, but you must always study everything closely, in order to find out why your subconscious intelligences send you your dreadful needs and feelings.

You have to use your thinking, either intuitive or intelligent, if you are capable of these, since these are specific to natural living beings. While natural living beings are capable of intuitive and intelligent reasoning because they are in themselves cognitive systems made of zillions of inner intelligences doing this kind of thinking themselves, for these specific circumstances. Therefore, we must understand both the intuitive thinking and the intelligent reasoning, only to be able to continue our mental model of a created living being, along with our mental model of life.

Cells, invertebrates, vertebrates, and humans are capable of intuitive thinking. Intuitive thinking consists of memories that have feelings embedded in them, made possible by the entire midbrain. However, all cells can do the same, only at a smaller scale, because all cellular components interconnect throughout their entire activity of fulfilling all specialized tasks within the cell, while all cellular components and therefore all cellular intelligences send themselves needs and feelings to help each other by specialization throughout the fulfillment of all specialized tasks within the cell, achieving everything in a flawless manner.

Notice the continuous similarity between cells and the entire organism, because within the entire organism, all cells are specialized and they cooperate continuously in a specialized manner through the same needs and feelings enhancing their continuous interconnectivity exactly as it is the case within cells, because the same cellular intelligences interconnect by the zillions while forming very large systems of intelligences spanning the organism, further interconnecting in the human society while forming entire societies tending to everything necessary in the entire outside world, while sending the same needs and feelings to perform the same specialized tasks in all forms of life and in all classes of life, from the bottom of the electromagnetic field all the way up to Life herself. Yet this is the case only in the electromagnetic type of life, because it might be different throughout other types of life forming Life and the wider world alongside the electromagnetic type of life.

The same intelligences, interconnecting continuously in a similar manner in the ionic form of life, molecular form of life, cellular form of life, organic form of life, social class of life, Earth ecosystem, and Life altogether, in this order, because the lower forms of life give life to the higher ones up to Life herself, always through needs, feelings, meanings, and specializations.

Therefore, the entire intuitive interconnectivity taking place at subcellular level, cellular level, organic level, social level, ecosystem level, and at the supreme level of the wider world is called intuition stereotypically, while intuition itself is very complex, made possible even flawlessly through the very well coordinated effort coming from zillions of living beings and intelligences interconnected flawlessly by specialization.

In the electromagnetic type of life, it was possible to live life only at the base of the electromagnetic field as living intelligent ions of the ionic form of life. The entire molecular form of life is an invention of the ionic form of life, an extraordinary achievement allowing living beings and intelligences additional space, niches, resources, and opportunities to live an entire expanded life right on top of the ionic form of life, with all living beings and intelligences of the ionic form of life making possible continuously the entire molecular form of life from the base up, while giving it life from its own life.

Furthermore, the molecular form of life invented or made possible the entire cellular form of life, with a drastic change of the environment that made everything toxic, while the molecular form of life had to protect itself in larger communities living life behind a very strong wall, the cellular membrane, while making all cells possible along with the entire cellular form of life.

Furthermore, the entire cellular form of life made possible the organic form of life and then the social class of life, the Earth ecosystem, up to Life herself, yet at a closer study, you notice how it is always the ionic form of life making all these possible while living life as all the other forms of life and classes of life above, as though the entire life takes place only in the raw electromagnetic field, only at its base, with all the other forms of life and classes of life above as simple expansions of the ionic form of life and of the continuous living intelligent encoded oscillation in the electromagnetic field. Life is the entirety of this continuously successful living intelligent meaningful specialized interconnectivity of all living beings and intelligences by the zillions, while this is the definition of life. We still have to model the living intelligent electromagnetic oscillation taking place at the base of the electromagnetic field making the entire electromagnetic type of life possible, yet at least we have the definition of life.

At the second animal intuitive level, you always have a coupled memory consisting of a solution – feeling, or knowledge – feeling, always acting together, because at the second animal intuitive level, all memories are coupled with feelings. Animal living beings do not exactly have the clear, distinct memories and knowledge that humans do intelligently, based on distinct accurate intelligent elemental concepts, since you have to use your intelligent reasoning to digest cognitively everything that you know about the world, in these clear elaborated discrete intelligent concepts. At the animal intuitive level, you understand and memorize the world only intuitively, through these coupled cognitive elements knowledge – feelings or memory – feelings. In this manner, at the second animal intuitive developmental level, you act according to your feelings, since the knowledge follows directly, because feelings are coupled rigidly with knowledge while leading the entire activity.

This is how you never place your hand in a hole in the ground filled with snakes, scorpions, and tarantulas, because you already have in your mind a dreadful feeling attached to your knowledge and experience with insects, in a coupled memory. Similarly, you always eat apples, chicken, potatoes, and bananas instead of dirt, branches, and leafs, because of all your coupled memories involving food, while always seeking to feel good and not bad. While at the third intelligent human level, you use third level conceptual reasoning throughout your safety and nutrition, alongside your second level intuitive thinking, and alongside your first level algorithmic reflexive thinking, in entire comprehensive mental models.

Which one is easier? Which one assures your survival, subsistence and development? How can we ever include all these within the rudimentary, mechanical cognition of our robots in order to render them alive? We cannot, because the rudimentary mechanical brain that our robots have currently in our model of a created living being is capable only of first level algorithmic thinking, as all current computers offer, by using a limited number of solutions for a limited number of circumstances, while all living environments are unlimited in events and possibilities.

There are advantages and disadvantages for both second level intuitive thinking and third level rational, intelligent, comprehensive reasoning. At the second animal intuitive level, by having your feelings coupled with your knowledge, you can use your coupled feeling – knowledge instantly, through simple reflexes. This saves your life, because intuition and reflexes are very fast, and they manifest instantly. Furthermore, you can transfer intuitive knowledge from one generation to another and from one species to another, as it always happens, because intuition is innate and conditioned. While third level reasoning is different, since it uses distinct, well-defined concepts, as cognitive Lego pieces that you place together throughout your elaborate intelligent mental models, matching in this manner entire lines and lifelines of causality, making you very successful.

Do not discard your second level intuition, since third level intelligent conceptual reasoning, even though is very accurate and very precise, is not too fast, and it does not occur unconsciously or subconsciously. Many of your inner intelligences still use it, and you have to remain capable to accommodate it throughout your inner interconnectivity within your cognitive system and in the outside world. It is the same with your first level algorithmic ideological thinking, since if you do not use this throughout your comprehensive reasoning, you might remain incapable to identify and acknowledge beliefs, stereotypes, and entire ideologies, since these are of the first consensual cognitive level. You can also miss your logic throughout your cognitive tasks, since logic and the entire discrete mathematics are of the first algorithmic level.

As a reference, the entire computer science is of the first algorithmic level, as it uses logic and discrete mathematics, while working hard to achieve intuition, but never achieving it. It is similar with us and with our robots, because we fail making our robots intuitive and or intelligent. It is impossible to make computers intuitive at the second animal intuitive level, because you have to tap into the living intelligences of the electromagnetic field to do so, since only these are intuitive. They are already intuitive or even rational and intelligent.

As a reference, when so-called aliens abduct people, they take small samples of their brain, they do so usually through the nose, which is painful and frustrating, but these groups of neurons that are taken already contain very valuable intelligences within, many times having highly advanced cognitive abilities, even superhuman abilities, genetic remnants from once highly successful and highly developed past civilizations of Earth, the old successful golden human ages. Furthermore, it is easy to graft these capable intelligences within other brains of living beings or within artificial computers, as part of a brain – computer android living being. These are intelligent, not only artificially intelligent, but naturally intelligent. Because if we have opted out from using living brains in our robots, the science of the upper social classes seem to use them well. At least this is what they were doing in the sixties and seventies while abducting people, because currently they grow and raise their own people and hybrids to do as they please.

Intuition takes place at conscious and subconscious levels, but mostly subconsciously, while science calls it simply reflex. This is how, with each decision that you have to take at this second intuitive developmental level, you already have the specific feelings embedded in your memory, letting you know in advance how it will feel after implementing your decision. In this manner, you can follow only your feelings throughout life, your gut feelings, your intuition, and you do everything, you achieve everything in life in this manner, but only at the second animal level. Which is better than the zero addicted level or the first servitude, ideological, consensual, algorithmic level.

For example, as an animal living being, you see a snake on the grass, and you have to take a decision fast, intuitive, intelligent, or by reflex, depending on your species, brain type, and general mood of the day, of going around the snake, eating the snake, stepping on the snake, ignoring the snake, touching the snake, moving the snake out of your way, playing with the snake, picking up the snake, or running away fast while screaming down the hall, since you always have a multitude of decisions to take continuously throughout life, while you always have to choose the best, adding to your fulfillment or failure. You have to do so fast, mostly directly through instant reflexes coming from the basal ganglia, or through intuition from the midbrain. If you trust your intuition, since it is only of the second animal level, yet it is still fast and capable. Yet if you are an animal on the pasture, this is the only cognition that you have, basic first level reflex, and second level animal intuition. This is how you react through both combined, and it saves your life.

Notice how, while reading the above paragraph about the snake, your intuitive thinking had responded promptly and accurately with each choice that you had read about the snake, sending you well in advance the precise need, feelings, problems, or achievements that you were going to have after implementing each one. It is important to identify your type of thinking, because if your response was instant, it came from the first brain, the basal ganglia. If your response was still fast, while including instant feelings, it came from the second brain, which is the reptilian brain, or the midbrain. While if you felt all these while still spending an extra fraction of a second to think intelligently while assessing the entire circumstance, everything came from the entire brain simultaneously, from the basal ganglia, midbrain, and cortex, in a reflexive, intuitive, and intelligent manner. However, it is more likely that you moved away very fast through your fast reflexes, and then afterwards you reacted intuitively at first, then intelligently afterwards, to study the snake well, since it was a unique opportunity.

The midbrain sent you your intuitive cognition, by coupling feelings with knowledge. You received this knowledge – feeling solution of how to overcome the problem with the snake as you read, sometimes agreeing with the solution that you were reading, and sometimes disagreeing, threatening you with the specific feelings that your primal security subconscious intelligences were going to punish you if you only picked up the snake, stepped on it, moved it out of the way, ate it, or just ignored it. It happens right now again, your security intelligences react accordingly, instantly, and flawlessly, right from the midbrain, sending you your knowledge – needs – feelings combined, in order for you to know exactly how to behave in the outside world while fulfilling your human needs for security.

These are your subconscious security intelligences from the midbrain, sending you your security needs, along with all knowledge, needs, and feelings related to everything that you must do in the outside world to save yourself and everybody else. You have other subconscious intelligences throughout the entire organism, tending in a specialized manner to the entire organism. However, if their specialization relates to the outside world, they monitor continuously the outside world for all significant details and circumstances of the outside world. Exactly as your subconscious intelligences monitored and detected the snake on the grass, and once they identified it, they sent you your needs, feelings, and instant reactions to save yourself and everybody else, through all three brains simultaneously, because this is why the organic form of life developed them, for you to use them in all circumstances.

Similar to your security subconscious intelligences, your eating or digesting subconscious intelligences monitor continuously the outside world for all possible food, sending you your needs to eat accordingly, depending on what specific nutrients you need. Similarly, your reproductive intelligence watches closely everybody of the opposite sex, focusing on the most capable ones, while sending you your reproductive needs continuously, along with all possible good feelings tempting you.

These are the actual raw intelligences of the electromagnetic field that we seek for our robots, helping them throughout all possible circumstances, exactly as they help all living beings of Life. However, these intelligences are the living beings themselves, because this is how all living beings are seen from cognitive perspectives, intelligences and entire systems of intelligences. While this is how all intelligences and systems of intelligences are seen from a physical perspective, living beings. Therefore, if we integrate the living intelligences of the electromagnetic type of life into our robots, we end up with the living beings of the organic form of life having robotic extensions, but not with robots having living intelligences.

You might assume that these are isolated instances of your

living mind and living cognition, but it is not the case. Because the entire mind and the entire organism are filled with these elemental living intelligences of the field, always standing at the base of all living beings and of all forms of life. How can you ever mimic and replicate these, while building yourself a living being? You cannot.

You cannot develop past the second intuitive level if you do not manage a comprehensive third level intelligent conceptual language, helping you understand and use concepts throughout reasoning and communication. While you have to be careful, in order not to confuse your intelligent concepts with beliefs, stereotypes, and strong personal convictions, since it downgrades your entire cognition to the first consensual, ideological, hierarchic, servitude level.

The only way to make sure is by using only accurate intelligent concepts, since these link directly with the natural laws of the universe, which include mathematics, classical physics, relativity, and firsthand knowledge and experiences, since these are real and many times alive. Accurate concepts are abstract, yet throughout intelligent human learning, you can attach them directly to the natural laws of the universe, as they are defined in an abstract manner by Existence throughout our spacetime continuum, present by default throughout real lines and lifelines of causality.

More precisely, while the second level intuitive reasoning uses feelings coupled with memories, the third level intelligent reasoning uses concepts coupled with memories and with feelings. Yet everything is more complex, since these intuitive coupled feeling – memories are not simple information, but they are genuine living intelligences of your cognitive system, living life in form of living conceptions throughout your conscious mind, intuitive and intelligent, in your midbrain and in your cortex.

Your conceptions are normal intelligences, only that they are not innate as it is the case with your subconscious intelligences, but they are formed, achieved, or conceived, since you always conceive them as memories while you learn and while you experience everything in the outside world. Your conceptions and all your intelligences relating to the outside world remain aware continuously of your environment, just waiting for the right circumstance to participate in your cognition or in the outside world alongside you the conscious intelligences. They just sit there in the background of your cognition, watching your cognition and the outside world live, and they wait. Then at the right moment, they act fast, and they can even take control of the entire mind or body in order to act and react exactly as they are supposed to do.

If these are the subconscious intelligences stated above, then you react exactly as they demand through your continuous needs and feelings, and you pick up the ripe apple from the tree while hiking to eat it immediately, or you protect yourself from the snake as we had already seen.

However, if these intelligences are actually achieved third level conscious conceptions from your cortex, you use them throughout your intelligent reasoning whenever needed, as it is the case right now while reading this book and while reasoning intelligently in parallel with this book, because this book requires knowledge about life and about a multitude of additional topics as reality, existence, interconnectivity, safety, nutrition, cognition, survival, perception, and wider world. All these are still considered intelligent memories or intelligent knowledge, you had formed them yourself with everything that you had learned about everything that you had found interesting and therefore important, yet these are not simple data in your head, but these are living third level intelligent conceptions living life normally alongside you the conscious intelligence in your entire conscious intelligent mind spanning the cortex, which is a distinct inner mind reality of the human mind, your most important mind reality as a conscious intelligence, because it allows intelligent reasoning, which is very powerful and very successful.

Within this inner mind world, all your conceptions monitor the outside world and the entire cognition for the most proper moment to act alongside the other conceptions as these form larger concepts and conceptions themselves throughout elaboration, as it is the case now while forming, elaborating, and further conceiving this large third level intelligent conception of life. This is the actual life, regardless if it takes place within mind worlds, in the outside world, throughout co-created classconscious realities, or in the higher worlds of the wider world, and this is what we must consider in our definition of life.

This is how you retract your arm fast from the hot stove, you jump away from the snake, you eat the apples but not the branches, you close your eyes fast, you prefer calmer people in your entourage, you learn about life, or you avoid all treacherous parts of your city and now nobody goes there anymore, because you simply had a feeling about it, it was all in your guts, in your intuition and in your intelligence, or this is what you tell everyone, while all these are the raw, living, elemental intelligences of the electromagnetic field standing at the base of your entire cognition, they are always alive, they consist you entirely from inside out, while through their lives, they give you life and they make you alive.

However, if you are a human being on that pasture, you might be tempted to reason diligently, which is slower but very pertinent, in order to learn everything about the snake. This is the case if you happen to be allowed to take your time to reason, since your security intelligences use intuition, and constrain you through all your reflexes to react immediately throughout all critical circumstances. This is the case if you manage not to get in their way throughout life as a conscious intelligence, since the harmony of your inner cognitive interconnectivity is always meaningful, and you must always maintain it. It is important to identify and study your needs and feelings in order to learn everything about all intelligences of your cognitive system, conscious and subconscious, while in this manner, you can interconnect positively and harmoniously with them within your cognitive system and in the outside world, since they are also found in all those around, and you must maintain harmony similarly with them.

Life

Because as seen, you live your life within all your environments, worlds, and realities simultaneously, while many times, you are at the center of your environment, and through yourself, you maintain harmony, cooperation, and therefore a positive interconnectivity within all your environments, worlds, and realities, within your cognitive system as a conscious intelligence, at home in the family as a living human being, at work, at school, in society, online, within hierarchies, jurisdictions, and ideologies as a corporation, or throughout your higher worlds and realities as a higher self.

You must interconnect harmoniously within your cognitive system with all your subconscious and inner conscious intelligences, while interconnecting harmoniously with the conscious, subconscious, and highconscious intelligences of all those around, of all your loved ones, exactly as you interconnect with your own intelligences. You must develop all these intelligences, or you must help them develop themselves by providing everything necessary exactly as they send you your needs, while fulfilling and developing the conscious, subconscious, and highconscious intelligences of all those around, since you interconnect directly with them and with all their intelligences, many times directly through all your intelligences, conscious, subconscious, and highconscious. Life itself is this highly complex and highly meaningful natural living intelligent fulfilling interconnectivity.

It is easy to understand and to develop intuitive thinking, because all that you have to do is couple feelings with memories or decisions, as it happens in all reflexes, innate and conditional. Yet you already do so unconsciously, through all intelligences of your cognitive system, since they are intuitive, while it is meaningful to interconnect positively with them while understanding them well, since you are the conscious intelligence. Since this is your task as a conscious intelligence, the accurate coordination of the entire organism throughout the outside world while fulfilling its needs. Needs than now become your needs, since as a conscious intelligence, you have the most pertinence with the outside world, and this is how you are allowed to overrule the needs, feelings, and intentions of all your intelligences in everything regarding the outside world. While this certainly adds to your failure or fulfillment throughout life, and it is important to know it.

This is why cognitive systems use feelings, only to help you predict events and circumstances, and therefore to help you make your best decisions. While you know exactly how to couple the perfect feelings with the specific knowledge, solutions, or memories, and you do so through experience, which can be your own, shared, or genetically embedded in your cognition from the previous generations.

You have your own feelings embedded in all your memories involving snakes, and you will certainly act accordingly. As an animal, you will always go around the snake while averting the others if they are not already aware, or you simply observe the snake in order to learn more, to create new intuitive and rational knowledge and memories, since curiosity alone demands it persistently.

This is your developmental need at an animal level, curiosity, when you remain successful continuously throughout life, since curiosity alone is capable to develop you to the level where you can remain successful in the fulfillment of all your needs. If not, the dreadful feeling of boredom and regret will constrain you to learn just as well, if anything terrible happens regarding your encounter with the snake. You have to follow it, building on your experience, while in this manner, as an animal, you certainly seek to eat as much grass around the snake, or to hunt it in the safest manner as a carnivore.

However, you also have innate memories and entire reflexes concerning snakes, spiders, scorpions, authorities, criminals, prosecutors, jails, sentences, divorces, bankruptcy, and low grades in school. You notice how your security intelligences reacted again, since you always avoid these through the same bad feelings embedded in them, many times through intuition alone. You do so through the algorithmic, ideological, judicial thinking of the first level, by following specific lists of beliefs, stereotypes, laws, and orders, or by

Life

following intelligent reasoning altogether at your third developmental level, including algorithmic, conceptual, intuitive, and intelligent thinking, and much more, in a consistent manner.

Reasoning takes place mostly through mental models, which are performed by a multitude of inner intelligences, while these live their lives normally, within inner mind worlds, which resemble the outside world. You as a conscious intelligence create your own inner replica of the outside world through learning and through the experience that you undergo throughout life. You perform some of these mental models yourself throughout your reasoning, in order to find solutions to all your problems. Your inner intelligences can perform similar models in parallel with you, helping out as much as they can, and they are the ones sending you the multitude of ideas that simply pop up in your mind and you use them as being your own.

This is how you are supposed to prevent any drastic major environmental event throughout life, if you are capable to perform mental models at such an advanced level. In order for you to be able to reason intelligently through advanced mental models, you must have an advanced, accurate intelligent inner mind world filled with all knowledge of the outside world, which is an intelligent inner replica of the outside world, based on accurate knowledge, but not on stereotypes, speculations, or beliefs, since these might be wrong, and your success and survival are compromised. This is the case with everybody, since people think mostly through speculations, beliefs, and stereotypes, while still calling it reasoning, ending up addicted, tyrannical, or enslaved in the outside world, while ruining the entire outside world.

Can we offer all these to our robots? No, because your pertinent intelligences performing advanced mental models right now as you read this book have a zillion inner intelligences on their own thinking in parallel, spanning throughout your mind one inner cognitive reality after another, down to the field itself, and through the field, down to the upper reality itself, where a multitude of higher beings struggle with the same cognitive topic, how to survive the drought, the extreme cold, the snake, the flooding of all crops, the hurricane, the invasion, the genocide, or the inevitable discrimination.

Can we supply our robots with all these, in order to render them capable to survive and live a normal life? No, unless we tap somehow into Life herself, to provide us with all these living, natural inner intelligences, since they are her inner intelligences, and we cannot replicate them algorithmically, since it is not enough.

You can study right now the multitude of your inner intelligences, to find them taking the side of Life, through all their needs and feelings, but never your side. While they will always kill you and get you out of the way if you ever go against Life, or if you only ignore or neglect Life, regardless if this also ends their existence. This is how you get sick and you never recover, because your own intelligences will never recover you, letting you die away, only not to harm Life anymore. You start suddenly experiencing very bad luck and it never ends, until you die, because your own intelligences will get in your way and will sabotage you continuously, until you die. You produce your own free radicals in large amounts and these take you apart one cellular component after another from the inside out until you die, and you never know why, which happens often. How often? Every time, because this is what happens with everybody at an old age, you produce free radicals in all your cells in large amounts, these take you apart from the inside out, and you die. Coincidentally, this happens exactly when you stop working for Life, when you retire from social and family work, and when you just take it easy. Death is an invention of life meant to remove continuously the old, the unsuccessful, the unfulfilling, the meaningless, the obsolete, the idle, and even the less developing. While in general, if you only blink suspiciously inadequately and less fulfilling, Life and all your intelligences take you out. While if this not enough, Life takes out entire organisms, families, genetic lines, nations, species, worlds, and realities, for similar lack of fulfillment.

There are many capable natural living beings and civilizations up there and out there seeking to create life on their own, they create life exactly as we do in this model through mechanical devices, computers, and artificial, algorithmic, judicial, or ideological intelligences that are only of the first level, and if they cannot manage to tap into Life herself, they can never include the specific essence of life into their artificial devices, and therefore they cannot have life, without exploiting you. More precisely, they exploit you, while you give them life, through your own life.

Yet you can always avoid them, because within you, your needs for freedom and security keep you away from tyrants and profiteers, while helping you save an entire world from them, because this is what your security, developmental, and social intelligences want, freedom, harmony, meaning, success, and fulfillment. Just follow your natural needs, meanings, and feelings, and you can find Life and the Divine through your own life, because you have them within. Because only naturally, you can be part of Life and the Divine, and it takes us this entire model with little robots, minds, intelligences, forms of life, matrices, needs, and fulfillment to figure it out.

Our robots do not have the full status of living beings. In theory, they have the chance to ascend to a second level living status from their first level mechanical one, if the environment does not remove them in the meantime. We hope that they can still develop, if the environment only gives them a chance, since science still promotes evolution through chance. While we already know that there is more than chance in this evolution that we call development, there have to be living developmental intelligences always demanding and even enforcing development whenever Life and the environment require it, yet we do not have direct access to these living intelligences in our model, since if we did, our robots were already alive and they were already developing even more.

Why do our robots survive so far, while they are not yet intuitive, and they are not able to cope with all changes in the environment? With each drastic change in the environment, not all robots survive, yet some or most of them do, since the environment strikes randomly, and most of the time, only locally. It is just a matter of chance which robot lives and which dies. Once a cataclysm takes place, solutions of survival are immediately downloaded to all unaffected robots from all lucky survivors, and therefore everyone becomes adapted to the environment, until the next strike.

This is intuition at its lowest level, since intuition is found in the power of many to think and find solutions on the same theme, and then to share the knowledge, elaborate it more, and use it throughout further thinking. This is why you need interconnectivity alongside intelligence and the field to have life, because only in the power and specialized pertinence of many, you can have diverse, new, original, successful thinking. This is why life folds upon herself while forming newer and newer forms of life that include trillions of individuals at a time, since its cognition can become trillions of times more complex and more diverse through all upper forms of life and through all upper classes of life. This living intelligent diversity is capable to create ideas and successful solutions throughout algorithmic, intuitive, and intelligent thinking, while helping life survive more unexpected circumstances but not all, since success is never guaranteed below the tenth supreme level.

We do not have in our model entire cognitive systems full of intelligences to think and survive on the same theme, but we have only a community of robots, each one equipped with one mechanical intelligence, always seeking successful solutions within a continuously changing environment. The robots survive significantly less than the natural living beings do, yet they still survive, so far, in this specific iteration. They survive by some measures that we have preprogrammed ourselves, along with some measures that they had managed to derive and adapt themselves from ours, as finding how not to slide on mud and on wet grass, from their previous solutions of not to slide on snow and sand. This might be expected, yet since they managed it, it is a success.

This is the case with all germs from the organic for of life, since the cures to all illnesses are always broadcasted immediately in form of messenger proteins by all surviving and recovering cells to all cells within the organism and outside the organism. All cells know how to cope with the specific germ, the entire organism is saved, and then the entire organism broadcasts the curing solution everywhere to everybody.

Why and how did our surviving robots have exactly the necessary solutions at their disposal? How exactly can a brand-new solution enter their set of preloaded solutions, since the robots are not able to think and learn intelligently or intuitively? According to the theory of evolution, success comes from errors, always randomly, involving chance. According to the theory of evolution, one of the solutions for a different environmental problem was damaged by accident, mutating exactly into the right solution to cope to a brand-new environmental event, while saving in this manner our entire community of robots. You can never believe that this is actually possible, but there is even a probability associated with this random process to happen, and it might be equal with your own probability to win the lottery, to fall off a plane, or to marry a multibillionaire. Which still happens, because people still win the lottery, fall off airplanes, and marry billionaires. Since here is the power of the many throughout ingenious thinking, and this is why chance might actually save our robots many times. Yet when they fail and when they are destroyed entirely, we simply rerun the model, to have our robots once again in a new iteration. Yet intuition, intelligence, and persistence govern life, not chance and probability.

If this world is created similarly, run and rerun throughout similar iterations, there is not too much need for intuition and intelligence throughout life and throughout the world, and there is no need for life altogether, since chance alone keeps everybody successful eventually. We are reborn from our own misfortune, with yet another rerun of the entire world, of this entire model we call Reality, ready to do everything all over again, hopefully successfully this time. Yet this is not the case in the outside world, but only in the inner mind worlds, one mental model after another, until you find the successful solution. While even throughout your mental models, you always involve your intuition, intelligence, and persistence fully, not random accidents giving your final result. Even at school you must always involve your intelligence and your continuous persistence in order to find the right answer, not random accidents giving you the exact solution in a miraculous manner.

Are chance and probability governing the world, as science states in many occasions and through many theories? There is a chance for everything in the world to happen, as slight as possible, even for entire species to survive, evolve, or go extinct, by chance alone. People compare this specific probability with the probability that a tornado can produce a brand-new jumbo jet by assembling randomly everything from a junkyard, ready for flight with all the lights blinking, while the chance for this to happen is still there according to the current science, as slight as it can be. It is possible, you must wait for some time to happen, but the probability is there according to the current science, so just wait patiently, and it will happen. It is certainly unlikely, yet this is the only evolution and adaptability that our robots can experience according to the current consensual science, chance and accidents.

The theory of evolution and the theory of survival of the fittest seem to apply to our little robots at this specific point in our model of a created living being, because the theory of evolution and the theory of survival of the fittest are made for artificial, consensual, mechanical beings similar to our robots, but not for living beings, not for life. Study Biology closely, to see how it always explains life in an automatic mechanical manner at the first algorithmic level, as though it tries to model exactly our little robots, not the living beings of an entire world. While they have never heard of intelligences, always considering only the physical bodies and the physical brain.

The current science states persistently that life itself evolves from the nonliving, exactly as we are trying right now to build an entire living being, while we never succeed, yet still making

Life

our model of a created living being very pertinent according to the current science, while never accurate according to Life and the real world.

We never accept chance in our model, not here and not in every model of this book series "Human," because if chance alone stands at the base of everything, then we do not have to model anything anymore, since we already have the answer: chance. Yet Life is never by chance, since Life stands above chance altogether. Life conquers chance whenever entropy becomes too high, and if Life ever uses chance, she does so to remove beings that are already inapt, disconnected, and incompatible. Interconnectivity is not by chance, since everyone and everything works hard throughout life within packs, herds, cells, organisms, and entire societies, in a combined effort that can become very consistent, but not at all random or by chance. Yet chance can still manifest, whenever interconnectivity fails, saving or destroying everybody in a random manner, while never helping us learn anything, never developing us.

Existence itself does not define anything by chance, but through very precise lines of causality, matching the lines of reasoning of Intelligence, matching the lifelines of achievement of Life, matching the lines of status, outcome, harmony, and competition of all higher forms of life, while even matching the very well ordered hierarchic lines of agreement of the Consensual Matrix. We are in Life, in reality, and in the entire existence, and nothing takes place by chance, because everything is interconnected directly or implicitly, but always rigidly, accurately, and purposefully, similar to all wheels of a mechanical clock spinning never randomly, but always connected. There is still chance throughout life, locally and shortly, but only at the individual level sometimes, and only while involving their failure and incompatibility.

Chance relates with errors and malfunction, but not with achievement and therefore development. Because development at the level of living beings is naturally embedded in themselves and in their entire form of life, while this is very well defined, part of the entire interconnectivity, inner and outer, while you either follow it, develop, succeed, and therefore you fulfill Life, or you fail and perish, and you get out of the way. With existence ceasing to define you well before you are gone, ever since you had failed developing, ever since you had failed Life, since you cease existing for Life once you drop outside of the meanings of Life. Therefore, nothing takes place by chance for Life, since you are either with Life, engaged in her entire well-defined living interconnectivity, otherwise her existence cannot define you, keeping you out. You can follow chance throughout life, but you might go against Life if you do so, because Life wants your continuous success and your continuous development, not only random or by chance, otherwise, eventually, she takes you out.

Why do we see chance promoted in all movies, music, and scientific theories? Chance is in many ideologies, yet even the Consensual Matrix avoids chance, since its entire bureaucracy is very well defined, and therefore there is no room for chance and error in all its jurisdictions. How many times do laws involve chance? Never, since all laws are always absolute throughout all jurisdictions and ideologies. How many times chance has priority in court? Never.

Why do they still promote chance in all entertainment and even in education, media, and science? Because education, media, and science are entertainment altogether, while you can find promoted in society everything causing people to divert, stagnate, decay, fail, miss opportunities, fall off their natural niche, get sick, lose everything, and in general, just clear the way and their natural niche, in order to hand in the world to all tyrants. Because once you start taking chances, you clear the way eventually, and others take the world, mostly tyrants. While with entire theories coming from science promoting chance, mostly through comprehensive theories as big bang and evolution that install themselves at the core of your inner replica of the world, this is how you promote failure in the world, through your implicit erroneous reasoning.

The theories of evolution and survival of the fittest as

offered by science are meant for mechanical, nonliving beings of the first developmental level, for automatic mechanisms, with the current science following us closely as we try to build our living robots, while we already know that we have failed our model of a created living being, since we cannot make our robots alive.

The current science cannot model life, intelligence, development, and thinking of all levels, and this is why they involve probability and statistics in their theories of life, but not natural needs, natural behavior, natural development, and living intelligences determining harmoniously all these. The current science cannot model intelligence either, at least not for the Masses, and this is why it cannot explain the development of life taking place naturally throughout individuals, species, forms of life, and ecosystems. Because once you can model second level and third level intelligence and thinking, you can model second level and third level development and interconnectivity, and this is life.

The current science avoids everything related to other realities, higher beings, higher laws, human development, human rights, natural human needs, and intelligences as normal living beings, because the current science is integral part of the Consensual Matrix, and therefore science cannot study life, while this might be the case only with the science meant for the Masses and Lower Brotherhood. Therefore, science refers to development as evolution, since the development and evolution of first level life are statistical and probabilistic. Science does so while attempting to explain what is observed in the species of Earth as a mechanical development, as legs growing longer, hoofs getting tougher, and heads elongating in time, all happening mechanically and lifelessly throughout life. Which is never accurate, because horses are second level life, and not first level mechanical or consensual beings. Yet since science cannot study life, science cannot explain horses, and therefore everything that it states is erroneous, while calling it a theory, only to state beforehand that everything is an assumption or a speculation, never accurate.

Life starts with the second animal intuitive level, while only our robots are of the first mechanical algorithmic level, along with all computer programs, current artificial intelligence, and machines and mechanisms of all kind found throughout plants and factories, along with you the consensual corporation, along with all brands and trademarks of the world, mega corporations, jurisdictions, districts, and consensual nations. These are consensual beings of the first level, along with the entire Consensual Matrix.

These are artificial and consensual forms of life, and from among all these, we try to develop natural life and intelligence of the second level and more, but we cannot. We attempt to do so throughout our model with robots in a scientific developmental manner, probabilistically, and it still works, as long as our robots can hang on to their mechanical life, as we still hope for the best.

Because everything is alive and intelligent in the world, and more importantly, everything is connected with everything in the world, and therefore everything is part of everything in the world, since this is how living beings survive, develop, adapt, and prosper together. Yet as simple mechanical toys, our created living beings have no chance to survive in this world, because if you ever want to make it alive in this world, you have to find a way to tap into the intelligences of the field, get yourself a zillion intelligences of the field within your own cognitive system, allow them to roam freely within their own specialized inner replicas of their environment in order to allow them to think even at very low levels, and this is life. Yet since our mechanical robots are doing the same thinking individually a zillion times fold, while keeping the best results, then this is still intuitive thinking, as slight and as mechanical as it might be.

You cannot manufacture your own living intelligences yourself in order to create life and intelligence in an artificial device, but you have to tap in the life that is already everywhere, at the base of this world, in the electromagnetic field. Therefore, we can never state that we can create life ourselves, yet we can still tap into it. Even the new intelligences created by the changes in the environment throughout the development of life were already in the field, and they only took the opportunity to take over the new cognitive niches produced by the new changes in the environment, developing themselves while developing in this manner the new organism to new meanings and new achievements. If the new changes are adequate for entire cells and for the entire organism, then these changes become available for the entire species and for all following species.

If new knowledge can be created by damaging the old knowledge as the theory of evolution states, then organic beings would make a priority in damaging deliberately and randomly their own stored information, only to 'diversify' their previous knowledge, in order to develop themselves. Yet this is never the case in life, in nature, and in the real world, while it never works in practice, since the organic form of life tends to strengthen as much as possible its integrity of knowledge, mostly when this genetic information is passed from generation to generation while making copies of copies endlessly, yet nothing is compromised.

Right now in our model of a created living being, if you agree with a mechanical, random, lifeless evolution of life as it is modeled by the current theory of evolution, which you should if you still watch Discovery, if you are still employed by science, and if you love your benefits and privileges in the current consensual hierarchic society, then our robots are always alive and always thriving, as mechanical and as random as they are. We can continue our model of life in this manner, randomly and mechanically, at the first consensual mechanical level of life, while watching our living robots obeying continuously the laws of probability and statistics, while evolving right into the proud faithful scientists of the Brotherhood and of the invisible kingdom, since they already match the Consensual Matrix at their first mechanical level. Yet our model of life would parallel the current science from now on and we never achieve to learn anything new and accurate after this entire effort, since the current science is inaccurate when it comes to life, intelligence, development, humanity, interconnectivity, intelligence, reality, cognition, and spirituality.

In contrast, if you do not agree with the current theory of evolution since it has only one observed statement anyway, that species evolve, which is only a theory or a speculation, as it is already stated in its title, then our robots break down irreparably right here and right now in our model, since they can never advance in development from their first level mechanical life to second level intuitive life, and they never achieve to become naturally alive, unable to cope with the environment by all current scientific beliefs. Which means that, unfortunately, we failed, since we cannot make our robots alive, because we cannot offer them a viable matrix of life similar to what Life has at its base.

Therefore, right now, life starts with the second animal level on our hierarchy of life, and not with the zero and first levels. The zero level is for the disabled, nonliving, and heavily addicted, everything that existence cannot define, since they do not interconnect with Life at their zero addicted nonexistent level. While the first level is for mechanisms, cattle, computer software, servitude, corporations, tyrants, ideologies, jurisdictions, hierarchies, and the entire Consensual Matrix.

Life starts at the second animal intuitive level, with all cells, cellular components, plants, and animals. Humans follow at the third intelligent level, along with all rationally intelligent living beings who manage an intelligent conceptual language throughout the entire interconnectivity, inner cognitive, and outer social. From the fourth superhuman level up, there are various higher living beings, entities, and intelligences counting in zillions, up to the tenth supreme level, which is the ultimate level of Life, the Divine, Intelligence, Consciousness, and Interconnectivity.

We still have to model the entire development of life, yet we already have the first traces of our mental model of life. Because once we tap into the multitude of natural intelligences of the field to integrate them and to allow them to live into our robots, if we only know how to do so since we cannot simply graft them there, then these intelligences can become of any developmental level, from zero to ten. Once these ascend to the second developmental level, then we have intuitive intelligence, and therefore we have intuitive life, as natural as it can be.

Notice how, not exactly our robots become alive, but these intelligences are already naturally alive in them, if we can ever develop them. This is the essence of life that we were so eager to find. However, you can already tell that this is not exactly an essence of life, but since it stands at the base of all living intelligences in the world, it is a source of life.

Had Life herself developed from the first nonliving level to the second living animal and plant level, exactly as the current science implies through its old theories? No, since Life is always alive, because Life is life. Yet are we ever capable to build a living being, by persisting to take it from its first mechanical level to the second animal level? No, since life is never spontaneous, but it is always there. Life is not held in this world as science implies, but life is this world altogether, with this world and with the entire wider world always part of her. Nonliving objects as clocks, computers, and robots cannot come to life eventually, ingeniously, or spontaneously, as the current science states and as everybody expects, because everything is already part of life, only in various levels of life, forms of life, types of life, and classes of life. While you cannot control these as you want, because they are very complex, with zillions of living beings and intelligences making them possible.

The only possibility to make our robots alive is by oscillating ourselves by hand all ions and all molecules consisting them in a living intelligent manner in the electromagnetic field in order to make possible all inner intelligences having their roots in the electromagnetic field, while mimicking in this manner Life herself. Yet even in this manner, the molecules forming our robots must be formed and charged in a very specific manner in order to become the specialized proteins and enzymes of the molecular form of life. We can take all molecules capable to sustain life from the grocery store, since they are in all meat, yet in this manner, we end up again creating a living being in Frankenstein style, while also using sawing machines. While there is no other matrix of life capable to maintain life in this world so successfully besides the electromagnetic matrix of life making possible the entire electromagnetic type of life on Earth, with the ionic form of life, molecular form of life, cellular form of life, organic form of life, and social class of life included.

However, this entire attempted mental model of a created living being is integral part of our larger, comprehensive mental model of life spanning this book, since it is the main topic of our book, life, while we already expected to fail creating our living being. We already know from the other books of this series "Human," that Life is everywhere, including everything. Because stereotypically, you might still assume that Life barely manages to subsist throughout a continuous living struggle in all worlds and realities, yet at a closer study, you notice how all worlds and realities are part of Life, created naturally directly by Life, throughout her comprehensive cognitive activity, since this is the Universal Mind.

Therefore, we cannot have a nonliving mechanical object at the beginning of our model of a created living being, since everything is alive continuously. Because this entire real world is integral part of Life, and everything is alive within, always shifting from one individual living being to another, from one ecosystem to another, from one form of life to another, and even from one type of life to another, while everyone and everything belongs to Life. This is why our robots become part of life every time we lose them, since they break down irreparably, and then they corrode and become part of living beings, because they are always part of Life. The nonliving cannot evolve into the living as science always claims, but everything is continuously alive, only in different forms of life and in different types of life, always shifting from one to another through life.

This stereotype is kept implemented by science through ignorance, integral part of the second law of thermodynamics, because entropy itself is erroneously discovered and defined by science. Because there is no entropy in the world, there is no disorder in the world, but only Life, and she is very harmonious, with everything alive.

The current entropy or disorder is considered by science only from the human perspective, not from the perspective of life, because from the perspective of life, nothing ever loses meaning and fulfillment in a living world, throughout a continuously constant entropy, making the second law of thermodynamics erroneous.

Your room and your entire house become messier in time if you do not clean them, yet it does so within life, by returning to its own natural state. You might not be able to see it happening at home, because your house is already in a human environment, the entire city, while always maintaining your entire perspective human. However, cottages are in the woods, in the natural environment, apart from the human perspective, and you can understand the second law of thermodynamics in its entire inaccuracy at the cottage. Speed up time, to see your cottage decaying substantially to a collapsed roof, rotten wood, and plants inside everywhere, even trees, allowing all life of the forest to thrive right there where your cottage was. For you it is certainly decay, loss, and disorder, yet for life is normal natural life throughout a continuously constant entropy. The second law of thermodynamics is erroneous.

By not understanding life and this world, science maintains an entire world undeveloped, struggling throughout the dark ages from one major ideology to another, under all possible tyrants.

There are countless of computer programmers throughout the world attempting to make their computer programs intuitive and intelligent, just by using very complex lines of code. Yet you cannot obtain intuition mechanically, algorithmically, computationally, or consensually, but you must tap into the intelligences of Life if you want intuition and intelligence, because both intuition and intelligence are alive. This is what the Consensual Matrix does, it taps into you to give it life, to give it your own intuitive and intelligent life, endlessly.

What is the difference between evolution and development? There is no difference according to the dictionary, and therefore we can use the two words similarly. There is a difference in the evolution used by science, since it involves chance, while development and natural evolution never involve chance, relating closely to all natural needs of all living beings. However, this difference is not in the concept of evolution, but in the scientific theory called the theory of evolution. Evolution is different than the theory of evolution, because theories are only speculations or assumptions, and they are fictional or consensual. The consensual itself is fictional, standing apart from reality and therefore from life, while only fiction is unreal and therefore nonliving, standing apart from life. The code of law is fictional, only a play, only an act, the act of law.

Can you simply take some living intelligences, stick them into an object, and you obtain life? No. We are defining and modeling life here through a physical body and a multitude of inner realities full of intelligences that are capable to take care of the physical body as it moves and interacts with its environment while fulfilling its needs and meanings. Yet all inner intelligences are already the physical body as seen from an objective perspective. While the physical body itself is all its inner intelligences, as the physical body is seen from cognitive perspectives. You cannot make the nonliving alive.

For example, in order for Life to make the human cortex, she did not take some random living tissue, injected it with some very capable third level intelligences, letting them loose in the cortex in order to evolve humans from the second animal intuitive level to the third intelligent human level. This is what we try to do with our robots, and we never succeed. Instead, Life took some skin, where conscious intelligences were already present because they were the skin itself, and placed it on top of the already present two brains as a cortex, since these intelligences were already the skin itself, while they were already specialized in intelligent conscious thinking. This is how you have a cortex on top of the midbrain that is on top of the basal ganglia, because Life did it before, by placing the midbrain on top of the basal ganglia in a similar manner.

Yet the above example is from a living perspective, to be easier for you to understand it, because all developmental intelligences had to develop a new brain capable to handle third level intelligent conceptions and therefore third level intelligent reasoning, and this is what they achieved. Intelligence was still possible in the intuitive reptilian midbrain yet in a limited manner, since the second midbrain had been developed longer ago, in order to accommodate second level intuitive conceptions always coupled with feelings.

As you study closely the midbrain, you find all glands capable to induce all the necessary feelings coupled directly to all memories, therefore forming intuition. While the entire brain has the structure of skin and cellular membrane, since all conscious intelligences are always skin and cellular membranes, because this is what they are physically, skin. The developmental intelligences seeking to develop third level conceptual intelligence took all intelligences from the base brain and midbrain capable to conceive, hold, and maintain third level intelligent conceptions, conceived them in large numbers on top of the midbrain, and allowed them there to interconnect with each other while becoming gradually capable to hold very complex third level intelligent conceptions, becoming in this manner the cortex that you know well. All third level intelligences were already skin from a physical perspective, and this is why the human cortex has a skin structure and appearance, as it is packed up inside the skull. This thick skin grows and expands in the skull throughout the development of organic life, wrinkling gradually with each new specialized conscious inner intelligence developing there.

Cognitive development takes place through a continuous intuition and or intelligence, by using intuitive and or intelligent mental models. You notice how tedious everything is, mostly as we work hard in understanding life herself by conceiving right now a third level intelligent conception of life while using a third level intelligent mental model taking the entire space of this book, which is the largest and the most complex third level intelligent conception that you will ever have, since life itself is omnipresent, omniscient, and omnipotent.

Which is an achievement indeed, coming from Life herself with her own developmental intelligences capable to conceive an entire cortex for all third level intelligent living beings including humans, furthermore coming from you as you manage to conceive this entire third level intelligent conception of life. Our little robots certainly help us along the way to understand life from their own outside perspective, yet do not expect them to grow metal hearts and metal ears randomly in the next billion years or so through the laws of statistics and probability while becoming genuinely alive as ironman himself, since it will never happen.

We will keep using our robots as they miniaturize gradually, while allowing us to reach the very small scale of the base of this world not directly, but gradually, through our continuously miniaturizing robots, while in this manner we hope to maintain a plausible nanoscopic perspective at the base of life, where everything starts. We would like to know how the ionic form of life manages to live life right on top of the electromagnetic matrix of life, in the raw electromagnetic field, in a continuously intelligent oscillatory motion in the electromagnetic field. We are also interested in how life manages to form newer forms of life on top of the old ones, and how all living beings and intelligences perform their entire living activity in a specialized meaningful manner by the zillions. How exactly do they know how to do everything in a specialized manner by the zillions?

It is easier to see how higher forms of life decay to lower forms of life through death. You can take a chicken, which is an organic form of life, kill it, clean it, package it, and place it in a grocery store, to have an actual transition from the organic form of life to the cellular form of life. Because all meat from the grocery store is of the cellular form of life, since it is still alive, but only at the cellular level. Yet even at the cellular level it is not in a sustainable state, decaying gradually to the molecular form of life. You go to the store, you buy it, you eat it, and then throughout digestion, all living chicken cells are broken apart into very useful living ions and amino acids of the ionic form of life, now fed to your cells gradually and methodically while offering them and your entire organism sustenance, prosperity, and development.

Life is lived primarily at the ionic form of life, while even the food that you eat must be broken down into ions and then it is reassembled within your own cells in all the cellular components that you know well. The food that you eat is not broken down into cells, because all cells secede. All cells are not born from scratch and they do not grow up from scratch, because as stated, all life is lived primarily at the ionic level, in the ionic form of life. While the organic form of life and the cellular form of life are only living mastodons formed and maintained continuously by the ionic form of life of the electromagnetic type of life. Coincidentally, the same ionic form of life found at the base of all living beings of Earth is present in stars, fire, and empty space, with all intelligences of the ionic form of life present in all these similarly and compatible everywhere.

In a matter of days, the entire chicken becomes you. More precisely, in a matter of days, you manage to shift life from the organic form of life to its cellular form of life and then to its molecular form of life throughout your digestive system, and then back to your organic form of life throughout your cells.

Which is the case with everything in the world, living and nonliving, shifting gradually throughout all forms of life and types of life, throughout longer or shorter periods depending of various circumstances. This happens not only on planets as this one, but in stars and empty space, since you can find Life everywhere throughout the wider world in all her types of life and forms of life and not only organic.

This is always hidden from you, the multidimensional life everywhere here and everywhere, with you included, and more importantly, with you always harmoniously part of life. Your meaning should always be with Life, and never consensually against her. Because this is why science makes a great effort to define only organic life as life, throughout all the definitions of life, in order to hide your continuous harmony with Life, along with your continuous harmonious meaning and fulfillment within Life.

While ignoring these, you remain eager to engage in all addictions, tyranny, and servitude in the world against the world, as it is the case with everybody in the world. People against people and life against life, one dark age after another. Because as you notice, only at your third intelligent level, you remain capable to fulfill Life harmoniously, and only through all these third level very complex intelligent conceptions about life, cognition, the world, and your place and meaning in life and in the world. What is life? What is harmony? What is intelligent cognition? What are humans? What are human conceptions, needs, feelings, development, meanings, attitudes, and intelligences? What are worlds and realities? Yet more importantly, what are all these combined forming life, with you caught right in the middle continuously?

Life can become highly complex from here in our mental model of life, with physical bodies holding inner realities filled with intelligences that hold inner realities filled with inner intelligences filled with intelligences in a multidimensional model of life, which is actually the case in the wider world. Notice how the realities of the wider world are made by intelligences and therefore they are part of Life, and this is how Life is the wider world, Intelligence, Interconnectivity, Consciousness, and the Supreme Living Being, as One. While there is no chance and even no error, but only intuitive, intelligent, and higher choice that all living beings and intelligences have throughout their life, development, and fulfillment, at least from the second animal level up, which are the natural intelligences of Life.

Life

This is part of the definition of life, and this is how our model with robots should be, if we can ever manage to make them alive. While as stated, you cannot simply take some natural intelligences from the field to inject them into an object to make it alive, because these intelligences must already be the object itself as seen from the inner perspective of their inner worlds, while the object itself must already be the multitude of intelligences and inner realities as seen from the outer perspective of the objective outside world. It is as a glove looking more as a glove from the outside and as a glove hole from the inside, while the inside and the outside of the glove are similarly correspondent. Intelligences must already be the physical body, and therefore we cannot harvest them from somewhere to place them into our robots, because they must already be in our robots, they must be our robots as seen from the inner perspective.

Because the matrix of Life is always the matrix of Life, and you can never create it nor produce it yourself apart from the matrix of Life. You can create and produce other matrices, as the Consensual Matrix and all computer worlds, but the matrix of Life is the matrix of Life, it is already there, and it cannot be otherwise. Which means that the nonliving cannot evolve into the living in any manner, randomly or not, it cannot be built into the living as we try here, and it cannot even be contorted into the living as the Consensual Matrix assumes that it does. The living is the living, in its own matrix of Life, as it is the Universal Mind, the One, and the wider world altogether.

By the end of this chapter, our robots will be used by Life within the matrix of Life, but only as basic ionic components digested within the ionic form of life, exactly as you have digested the living chicken yourself as part of the same ionic form of life. Life always takes our robots back systematically, digesting them through mechanical failure and then through basic corrosion, as she always does throughout this mental model of a created living being.

Which means that there is no cell of life, no bone of life, and no neurotransmitter of life giving your entire organism life, because you are alive from within, through everything that you already are. You are alive through the zillion living cells, cellular components, organs, and intelligences. Which means that there is no essence of life, because you are entirely alive, you are entirely the essence of life, the quintessence of life, and even the origin of life. Can anyone build this from scratch, apart from life and the matrix of life? No. Frankenstein succeeded, yet he used sawing machines within the matrix of life, which is not exactly creating life from scratch, but only modifying it.

Therefore, our nonliving robots must already be alive and intelligent before we even intend to start making them alive and intelligent out of simple plastic, glass, rubber, and metal, which is not the case with this entire project of creating a living being from scratch only to be able to discover the essence of life. The intelligences themselves are the physical body as seen from the inner perspective, and they must already be the body, which is not the case with our robots, because all intelligences live, exist, are intelligent, interconnect, develop, and therefore subsist, survive, and develop at the very base of all elementary particles that form the physical body, at the base of all ions, atoms, nuclei and electrons that form the physical body. This is why you have to digest the organic form of life and the cellular form of life into the ionic form of life to take it to your cells, since you have to go back down to the source of life with each meal, back down to charged ions and simple molecules oscillating in the raw electromagnetic field. While this is the Field or Universal Mind, since the Field is already alive and intelligent. This happens through the specific intelligent encoded manner in which these elementary particles vibrate in the field and therefore interconnect with each other, at the very base of the field, and it happens through the field.

Our mental model of a living being had already approached the core, hearth, origin, or quintessence of life in the above paragraphs, as these already include several supreme characteristics of Life or supreme laws of Life, as the supreme law of Mentalism, the supreme law of Correspondence, the supreme law of Vibration, the supreme law of Causality, and the supreme law of Harmony, holding directly the supreme law of Meaning and the supreme law of Development, which are supreme characteristics of Life, and this is the actual essence of life. Try to place these in our little robots if you can.

Because you cannot take some glue, plastic, and metal and hope that everything is alive and that everything starts moving around once you assemble it into a little robot, because life is more complex. How complex? As already seen, life folds upon itself to form newer and higher forms of life, and only in this manner, the intelligences of the field can manage to interconnect and inhabit at larger scales the entire macroscopic physical body composed by all its elementary particles. Which must happen naturally, since as seen, only Life is capable to fold upon herself to form larger and larger forms of life. As seen so far here on Earth, you have subnuclear life and intelligences interconnecting to form ionic forms of life. Ions can gather into molecules to form the molecular form of life. Proteins, RNA, and enzymes gather to form cells, and this is the cellular form of life. Then cells gather into organisms forming the organic form of life. While organisms should gather into living societies to form newer and newer classes of life and forms of life above, if the Consensual Matrix had not stopped it from interconnecting naturally and freely, killing it.

If we still want to build our living beings, we have to start from the lowest matrix of Life, from the basic field or Universal Mind, since the following matrices of life stand on this, and it cannot be otherwise. This is our last attempt to build a living being from the nonliving, as science always claims that it is possible. Yet this time, we will attempt to inject in our robots the essence of life, through all supreme laws and supreme characteristics of life stated above, while we have to do so all the way from the electromagnetic field, from the origins of life. This is only the origin of the electromagnetic type of life, since this is what we are, while we do not know anything else. Which is still fine, since by using the electromagnetic type of life, we might end up with a created intuitive living being as a chicken, so we get our chicken back. While if we really succeed, we might end up with a created intelligent living being, as a living human being, Frankenstein himself.

Notice how it takes several distinct forms of life only to go from the intelligences of the field to the subnuclear form of life, then to the nuclear form of life, then to the atomic form of life, then to the ionic form of life, then to the subcellular, cellular, organic, social, reality, and back up to Life herself.

Therefore, you cannot simply build a robot and expect it to become alive as Pinocchio from the story, since it takes accurate interconnectivities of intelligences on top of accurate interconnectivities on top of accurate interconnectivities to do so, one form of life on top of another, as far up as Life lasts. This is the definition of life, its actual model, as it includes life, intelligence, existence, the field, reality, consciousness, and interconnectivity as one. While you cannot build these in any manner in order to manufacture life, but these have to be alive from the start, while going through all these forms of life from the bottom origin of life in the field all the way up to Life herself.

There is no separate essence of life helping you build life, since life is everywhere, at all levels if it is intelligent and interconnected, or if not, life is still there, but only on very low forms of life, at the molecular, ionic, and subatomic levels. Yet it is the same life, with the same living intelligences still interconnecting in the field, but in lower forms of life. You can certainly consider an origin and a source of life, found continuously within everything, somewhere in the Field, and therefore always within yourself. This is why you cannot take some metal, plastic, and glass to form life, capable now to interact with the environment at your own macroscopic level, because it never happens, since the life already present in the metal, glass, and plastic can interconnect only at the ionic scale or lower, at the subatomic scale, but not above at higher scales and in higher forms of life matching the cellular and organic forms of life. You have to start from below, from the raw electromagnetic field, since this is the first matrix of Life.

Which means that, if you want to have life capable to interconnect at your own macroscopic scale, you must have a body with its intelligences already interconnecting at the macroscopic level, throughout a medium that allows them to do so. Therefore, this body is already alive, it is a living being, and therefore we do not have to build it anymore. While as we notice in the organic form of life, it takes a very long time for these intelligences to form larger and larger forms of life allowing them to interconnect at the macroscopic level within their own body, and through their physical body, to interconnect with the entire outside environment while forming themselves higher and higher forms of life, as far, as high, and as long as the wider world allows. Yet if you happen to be worthless enough to throw a consensual monkey wrench in all these, you kill the entire life, keeping it within very low forms of life continuously.

Each elementary particle, when studied minutely, is not material as you learn in high school, with those plastic balls and plastic sticks forming interesting patterns. There is more to particle physics, atomic physics, and nuclear physics, since theoretical physicists apply mostly quantum mechanics to particle physics, which is a probabilistic and statistical physical study, but not actual physics. While even electrodynamics is not sufficient to understand and model elementary particles along with the behavior of elementary particles. Even Maxwell's equations are not sufficient to model elementary particles, because you have to model the field not through the disciplines mentioned above, but through the interconnectivity of all elementary particles taking place in the field in a living manner, since everything is intelligently encoded and not only physical in nature, it is intelligent and therefore alive, marking the difference between inorganic chemistry and organic chemistry. Since this is why you have to use lifelines of causality throughout all your intelligent mental models, and not lines of causality. I refer to this entire living interconnectivity taking place in the field as being the raw intelligences of the field. While as stated, the elementary particles themselves are not rigid and material at their own level, but they are field distributions, vibrating lively and intelligently in the field.

While there is no statistics involved here, and therefore no quantum mechanics, since the entire motion of all elementary particles in the field is very precise. It is as precise as the entire binary language succeeding very rapidly and very complexly throughout your computer software while you play your favorite videogame. Everything is very precise, in order to allow your computer to make possible the entire inner artificial world of your videogame, otherwise you have computer errors, which never happen.

Now try to model these with Maxwell's equations if you can, while Maxwell's equations are your only option from the entire physics to model elementary particles and their comprehensive behavior in the field, which is very complex. Maxwell's equations can model only lines of causality in the field, which are of the first level, but not lifelines of causality, intuitive, intelligent, and higher. Nothing in the current science can model life, along with lifelines of causality, because the current science is of the first, algorithmic, consensual level, while Life starts with the second animal level. Your only option is to use third level intelligent conceptual mental models, as we use right now while studying life.

We cannot model Life past the third intelligent conceptual level, which is the human level, while Life goes up to the tenth supreme level. We will never understand Life at her tenth supreme level, since we lack the cognitive and interconnective means. While as you study the wider world, no living being and intelligence is at the tenth supreme level, besides Life herself. Life is the comprehensive life of all types of life, forms of life, and classes of life, from all worlds and realities.

Yet if we are capable to understand life from the raw field all the way up to the organic form of life and all living human beings at the third intelligent level, it is very good, because we can always match our own third level intelligent human cognitive capability, which is very good. We are lucky to have an intelligent cortex along with all the necessary intelligent knowledge to understand life while avoiding all beliefs, stereotypes, and entire ideologies meant to derail our intelligent reasoning. We are lucky to know exactly how to conceive, maintain, and conduct entire complex intelligent mental models to understand life throughout our reasoning, while using successfully all pertinent very complex third level intelligent conceptions related to the concept of life throughout this research, which is significantly more than whatever the current science does through its millions of consensual scientists throughout the world.

There is more to consider in the raw field, since the field itself manages to transcend through the matrices of all realities, interconnecting everywhere in the wider world, as the Field or Universal Mind, allowing all its raw intelligences to interconnect freely interdimensionally from one reality to the next while forming Intelligence. You have Existence defining these as being alive, since from an empirical perspective, you can define these to be alive, since Existence defines Life directly as being existentially alive, because Existence is the existence of Life. Because you never have distinct, individual living beings in the wider world, as rabbits, worms, and roses, but you have entire lifelines of existence, similar to the mind, body, and soul lifeline of existence typical for human beings here on Earth, yet it is the same with all intelligences.

Life is present implicitly in everything because life has been defined through empirical terms and words. Intelligence itself should take all the living credit instead of life or Life, which are there only as empirical supreme perspectives of Intelligence, Interconnectivity, and the wider world, since all these are supreme perspectives of each other. We have Intelligence as an intelligent supreme perspective, we also have the Interconnectivity supreme perspective, we have the wider world as the physical objective perspective, and we have the Deity as an ideological and spiritual supreme perspective. This is the definition of life as seen from supreme perspectives.

Which is which? These are the supreme perspective of what

exactly, if even the Divine is only a spiritual supreme perspective? These are the supreme perspectives of the Supreme Living Being, which is alive, conscious, objective, interconnected, and intelligent, as One. You can certainly consider that these are the supreme perspectives of the Divine when you base your considerations on ideological basis, yet ideologies alone do not consider all supreme perspectives of the Supreme Living Being. While all supreme perspectives are the same, each one defining each other from their own perspective, all defining each other individually, all being the Supreme Living Being. It is only through your limited ability as a living human being to understand everything comprehensively in all realities simultaneously, because you are bound to this world, since your conscious perception, awareness, and understanding are enclosed in your cognitive system, in the human cognition. You have your current ideological constraints including all scientific and social unfavorable conditions forcing you to understand supreme higher knowledge in this modular, truncated manner. This is why you have to dissociate the Supreme Living Being into Life, Consciousness, Intelligence, Interconnectivity, the One, the Divine, and the wider world. It is you dissociating through your limited perception, limited awareness, and limited understanding, but not them, not the supreme perspectives themselves, since these are always One. They are One Being, the Supreme Living Being. When you study each supreme perspective closely, they always diffuse one into another, into Oneness or Supreme Living Being, the way Life diffuses or transforms into Intelligence when studied closely, having even its essence in the raw intelligences of the field. While through the raw intelligences of the field, Life has its essence in the higher and higher intelligences found in all higher realities above, up to the supreme reality where Intelligence consists of all intelligences of the field, of all intelligences of the wider world, and therefore of Life. From supreme perspectives, all supreme perspectives transform one into another, since they are One Being.

Life

From supreme perspectives, we have to define life through all its supreme perspectives, because of our limited human condition and because all supreme perspectives are the same, they are One Being. Yet all these supreme perspectives are not exactly the perspectives of the One Being, but they are their own perspectives individually and through themselves, since Interconnectivity takes place in this specific manner. They do not have the Supreme Living Being at the top and they are not characteristics of the Supreme Living Being, but they are the Supreme Living Being simultaneously and comprehensively as One, while Existence defines all of them similarly.

We should not add the Consensual Matrix here, since it is not a supreme perspective, because the Consensual Matrix is not capable to reach any of the above. The Consensual Matrix strives to achieve consensual grounds, interfering with the natural Interconnectivity in this manner, and therefore through Interconnectivity, interfering with all supreme perspectives: Intelligence, the wider world, Life, Interconnectivity, and the Divine. The Consensual Matrix does not even span the entire wider world, being very limited itself, at its first developmental level, which is the level of our robots as they are right now in our mental model.

Why do I have to make mental models throughout my books? Because at the third intelligent developmental level, we understand everything through mental models of all levels: algorithmic mental models of the first level, intuitive mental models of the second level, intelligent conceptual mental models of the third level, and comprehensive mental models through all these at the same third level, up to the tenth level, which is the level of Life, Intelligence, wider world, Interconnectivity, and the Divine.

Mental models help us understand previous knowledge, they help us integrate, elaborate, and generate new knowledge from previous knowledge, through successful ideas and through cognitive interconnectivity. Most importantly, mental models help us expand our knowledge through any possible manner, circumstance, event, possibility, and correlation, by running the mental model or mental simulation in every case, every time we want. Mental models do not only help us understand everything while solving problems through the successful ideas that they generate, but mental models help us perceive, identify, and understand what is still not in our perception, awareness, and intelligence, reaching it just by running mental models in any circumstance we want. This is part of the human abilities, part of your common reasoning, if you keep it accurate and intelligent.

Our little robots form the mental model of a created living being, which is integral part of the comprehensive mental model of life, because life is the topic of this book. Our little robots stand somewhere on the side of our study of life, at its beginning, meant to help us understand, trace, and create the mental model of life itself spanning this entire book. Furthermore, our nonliving robots offer us an outside perspective of Life, since they are not alive, reaching knowledge and understanding beyond our current perception and understanding of life.

You can achieve intuition artificially if you can employ anything cognitive and animal in nature throughout your reasoning and throughout this study, because intuition is made of memories and feelings rigidly combined, helping you take decisions instantly, based on the feelings that your memories or previous solutions carry, by choosing what feels best. Which means that you have to be alive in order to be intuitive. However, intuition can be simulated through the interaction of a very large number of nonliving beings, even mechanical beings at their first algorithmic level, if their goal and characteristics happen to match the goal and characteristics of life, and if they are capable to learn one from another, forming in this manner a very slight interconnected mechanical matrix similar to the matrix of life, which it is still intuitive, as slight as it might be. Yet this is not life, but only a slight simulation of the second level intuitive life, made at the first algorithmic level, because it is only a simulation. We could use this in order to continue our mental model of a created living being while miniaturizing our robots down to the base of this world, in the raw electromagnetic field.

Our robots have an intrinsic goal to survive, subsist, learn, and develop, which is partly similar to all living beings, as all living beings are significantly more complex. This is not a living matrix of life, but a simulated correspondent mechanical interconnected matrix of the first algorithmic level, easily achieved by any computer program, yet never alive. It is similar with the consensual ideological juridical first level matrix of the Consensual Matrix, since this cannot become alive either, as it only exploits life in any possible manner.

Similarly, at its first consensual ideological juridical level, the current Consensual Matrix only mimics life, barely achieving this slight intuitive resemblance to Life. Yet all living beings composing the Consensual Matrix are alive, giving it their own lives continuously only to make the Consensual Matrix possible. While at a closer study, you notice how the Consensual Matrix does not use mechanical robots directly as we do in this mental model, but the Consensual Matrix uses intelligent living beings while constraining them to think, behave, and interconnect at the first consensual juridical ideological level, as though they are our little robots, and they always break down, mostly in the musical chairs scenario. In this manner, it takes most of the wider world to achieve intuitive cognition and intuitive interaction in the Consensual Matrix, as slight as it is.

We are not alone while seeking to build our own living matrix through our little robots offering intuitive cognition, intuitive interconnectivity, and therefore intuitive life, since all computer programmers search persistently a similar digital intuition to add to their artificial intelligence, hoping to advance cybernetics at the second intuitive level or even at the third intelligent level, while never understanding intuition and intelligence clearly, because the current science fails to understand them itself. The current computer programmers attempt to achieve artificial intuition and artificial intelligence by improving the first mechanical level computer hardware and the first algorithmic level computer code, which is always insufficient. From the first nonliving digital algorithmic computer level, you can advance to the second artificial intuitive level only through a continuous intuitive interaction between a very large number of artificial intelligences having the same goal: survival, subsistence, learning, and development. This is very similar to our robots.

Coincidentally, in cybernetics, this is also called robotics, mimicking life exactly as we attempt to do with our robots, and exactly as the entire Consensual Matrix does for a very long time with its own robotic incorporated intelligent living beings counting in zillions, with you included, because the dark ages never ended. This is why the computer programmers still persist to improve computer code while seeking to reach artificial intuition and artificial intelligence, instead of using interconnected robotic computer models, as we will do with our robots while miniaturizing them all the way down to the raw field. We will not reach intuitive life, but only a simulated version, which is still enough to help us understand the premises of the ionic form of life in the raw electromagnetic field at picoscopic and nanoscopic scales.

Notice how there is nothing rational and intelligent in intuition, but a simple reflex based on all your previous experiences, innate or achieved. Which means that cybernetics should aim by now not for second level artificial intuition, but for third level artificial intelligence. They might claim that they have already achieved it, yet they did not.

Any artificial intelligence, as any mechanical or digital computer intelligence, cannot give us natural life, but only mechanical life. This is not of the second intuitive level, while we do not even consider it alive. We cannot have life without living intuition, while we cannot have living intuition without life, because life, intelligence, and interconnectivity are not distinct components of life, but they are perspective of the same living oneness, defining the same living oneness. This is the essence of life, yet it is a source, not a type of intelligence, but it is Life altogether. You cannot create Life, but everything is already alive, only on different forms of life, all standing one on top of another, having their own source of life, called Life. Therefore, what we actually seek in our model of a created living being is more than an essence of life, more than a source of life, but we seek to interconnect the lower subnuclear or atomic form of life already present in the metal, plastic, glass, and rubber composing our created living beings, while helping it to interconnect furthermore into higher forms of life, passing or not through the cellular and organic forms of life, but arriving eventually up at the macroscopic level, where it can interconnect on its own within our current environment with the rest of the living beings, in a successful manner. This is what would make our robots alive, a successful living interconnectivity at a macroscopic scale, at our own level, which will never happen, because we cannot match Life herself. The Consensual Matrix tries similarly for a very long time, throughout most of the wider world, and still fails.

Life diffuses into interconnectivity now, after diffusing into intelligence first, because the three: life, intelligence, and interconnectivity, are one. More precisely, we always perceive *life* through our own objective perspective, but we find *intelligence* at a subjective, inner, mind, cognitive perspective, while we find *interconnectivity* from an even lower perspective, since intelligence is possible only through the intelligent, comprehensive interconnectivity of all inner intelligences composing it.

The zero level of life is nonliving altogether, along with the disabled, the highly addicted, the very sick, consisting all irrelevant helpless life, everything that does not contribute to Life in any manner, becoming the nonliving, since Existence cannot define it away from Life. The zero level is both the nonliving and the nonexistent, with everything consensual and the entire Consensual Matrix here at the zero level. Only that, by agreement, everything that is consensual, along with the entire Consensual Matrix, place themselves at the first nonliving level yet still existing, by agreement that they exist, as nonliving corporations or as unreal ideologies and jurisdictions.

While if you agree with them, then you and the entire Consensual Matrix are of the first consensual level but only according to yourselves. If not, then the Consensual Matrix is of the zero irrelevant level according to the rest of the world. However, the Consensual Matrix states by its own laws everything that it wants, even that it actually exists consensually, and now this is the first consensual level, according to them. Because for Life herself, the Consensual Matrix is always of the zero nonliving and inexistent level.

The Consensual Matrix can write any law that it wants, even that it is part of Life herself at the tenth consensual level, yet it has to obey higher laws, otherwise it risks to be dissolved by higher beings. Since by standing apart from Life and the real world, the Consensual Matrix avoids all responsibility within Life and the real world, doing everything as it wants.

The first level of life is the tyranny, servitude, ideological, algorithmic, digital, hierarchic level, everything that does not contribute to Life in itself directly, but it still helps Life through those who make use of it, as all tools, machines, robots, cars, computers, software, slaves, servants, followers, consensual beings, jurisdictions, orders, corporations, laws, and ideologies. Because the first level is still existent, but it is nonliving.

The second level life is the animal, intuitive, free life. All plants and animals are here, along with all cells, cellular components, packs, schools of fish, microbes, and viruses. This is the actual bottom level of living existent life, as it remains reflexive and intuitive in cognition.

The third level of life is the intelligent life, based on conceptual reasoning and comprehensive communication and interconnectivity, as it should be typical to all human beings. Yet it is not, since not all humans live life at this third intelligent level. Since you have made it so far in the book, you have been at the third intelligent level continuously, since this book keeps you at the third intelligent level in order for you to be able to elaborate it. Study closely this entire third intelligent developmental level as it manifests currently in your cognition and in the entire organism, through all its third level intelligent needs, feelings, modes of life, modes of cognition, cognitive interconnectivity, cognitive fulfillment, living fulfillment, family fulfillment, and social fulfillment, since you might not have the chance to do so later on when you do not read this book. Because this book keeps you at a third intelligent level continuously, as a continuous part of your third level intelligent human environment, as slight as it can be. With you developing this third level intelligent human environment furthermore, spreading into your family, entourage, learning environments, and working environments.

Several hours after you finish reading this book, you can witness the switch to the second physiological level whenever you have to, or to the first legal, consensual, ideological level if this is the case, or even to the zero addicted nonliving level if it ever happens. This is how you always live your life, switched continuously from one developmental level to another, from one mode of life to another, and from one environment to another, as it is important to remain continuously aware of this living detail for as long as you can. This is always the case throughout the dark ages, because during the golden human ages, you remain continuously at your third intelligent human level or higher, because this reality is of a higher existential level allowing cognition, interconnectivity, and therefore development past the third intelligent human level.

The entire human society was supposed to be at the third intelligent level, living its own social life at the third intelligent level, but it is artificial and consensual instead, at the zero addicted level and at the first consensual level, because you cannot have these two separate from each other. With everyone proud to keep it in this undeveloped dead manner for material interests, while this downgrades all living human beings from their potential third intelligent level of life to the second level animal life, to the first level consensual life, and to the zero level addicted vicious life.

Therefore, right now in our model of a created living being, if we want to create artificial living beings, we are already successful and our task is already over, since we have managed to create first level artificial, mechanical, robotic life, similar to your smartphone, cars, and drones. Yet if we want to create normal, natural artificial living beings, we have to create life at the second animal natural level and above, and therefore we have to continue the model, hoping to achieve all characteristics and abilities of second level normal natural life, including feelings, natural intelligence, awareness, learning, and development. We will not achieve these, yet at least we manage to miniaturize our robots to the nanoscopic level of the ionic form of life sitting right on top of the electromagnetic field, helping us understand it better.

How can we obtain intuition, in order to obtain second level natural life? We can do so naturally, not mechanically, neither artificially, nor digitally, nor consensually. What we know so far is that intelligences can easily perform algorithmic thinking of the first level, done easily by any artificial intelligence and computer program or software, since this is how computers function, algorithmically. However, it is through the intelligent interconnectivity within the field and within their inner realities that intelligences manage to achieve intuitive thinking. How exactly? It is very similar to our robots as they interconnect throughout their physical environment within their community. Robotics.

More precisely, it is not enough to improve only your computer hardware and software in order to achieve artificial intuition and artificial intelligence, because you must also assure the artificial interaction of a large number of interacting artificial intelligences, called robotics, while comprehensively, they could achieve higher artificial cognitive abilities, with the artificial intuition included. It is similar in life, because not only the physical body assures life, but also the intelligence of the physical body found in an inner reality, along with the entire interconnectivity of the physical body with a large number of other physical bodies towards a common goal. In this manner, you have intelligence, interconnectivity, and physical bodies being life, or being alive. Similarly, all intelligences interconnect lively within all mind worlds while tending to the entire cognition in a meaningful specialized manner.

Existence is of an objective nature in the specific inner reality where intelligences live, at their own level. From your upper perspective of this world, all intelligences of your cognitive systems are subjective in nature, performing all the necessary cognitive tasks and activities of your mind and of the entire organism, through their own thinking and thoughts. However, from the perspective of your intelligences within their own inner mind worlds, everything is objective in nature everywhere, they live a normal objective life wherever they are, and therefore everything that they manage to achiever throughout their normal life along their normal interconnectivity is normal, objective, and done by hand. While from our perspective, we call their acts and achievements algorithmic thinking, intuitive thinking, intelligent reasoning, thoughts, results, or ideas, and if these are useful in any manner, we call them successful results, successful procedures, and successful ideas. These are priceless achievements not only for our intelligences as they live their lives within their inner worlds, but for all of us up here in the world. We cannot get enough since all ideas are priceless, always helping us survive, subsist, develop, and prosper.

How exactly can you generate unlimited successful ideas? Through your intelligences and through their inner and outer interconnectivity, at their inner level within the human mind, and at the outer level in the outside world, called thinking. Thinking can come at many levels, including the second animal intuitive level, according to the level of their interconnectivity, and therefore determining the level of life, since all numbers of all levels are correspondent. You can find successful ideas on your own, as a conscious intelligence, throughout your comprehensive reasoning, which includes algorithmic, intuitive, conceptual, rational, mental modelling, and intelligent reasoning, since these can help you achieve everything you want, for as long as you are capable to develop them and employ them. If not, you have to think at the first ideological and algorithmic level through logic, laws, beliefs, orders, and ideologies. The current science is an ideology in itself, along with justice, religion, spirituality, and the entire society. If you are free of beliefs, stereotypes, and ideologies, you can use your own intuition throughout a free, animal thinking based on feelings, if you are capable to develop it. While if you know everything about the human mind, you become capable to develop it and use it comprehensively, through a comprehensive reasoning that includes the first level logic or algorithmic thinking, second level intuition, and third level intelligent conceptual mental modeling.

Therefore, while seeking to create a living being, we must focus on the second level natural life and higher, on the second level intuitive thinking and above, on the second level free, natural interconnectivity or more, and on the natural forms of life that include the natural intelligences of the field, only to be able to create a living being. While if we remain in the electromagnetic type of life, it is even better, since we can understand it more, seeing everything clearly exactly as it takes place from the bottom up of the electromagnetic type of life, starting with the ionic form of life.

As a reference, the natural second level interconnectivity and higher is capable to assure to living beings the necessary successful relationship among themselves, helping them fulfill all their natural needs and meanings, as it is the case with all wildlife throughout all ecosystems of nature including the jungle, but not in the current consensual society. Because in the current society, there are laws, norms, regulations, ideologies, vices, addictions, servitude, and material constraints interfering continuously with the normal, free, natural interconnectivity of all living human beings. This is how entire genetic lines die out and go extinct, because people cannot afford to have more than one or two children, or they do not afford to marry and have any children, or people prefer servitude and addictions instead, since the current human society constrains everything harmful to life.

Which means that the current human interconnectivity is of

the first servitude, consensual, hierarchic level, which is the level of its major social jurisdictions, laws, and ideologies constraining it, as capitalism, communism, nationalism, socialism, egoism, and sentientism of the Masses, or the egoism, sentientism, capitalism, communism, racism, and nationalism of the current consensual Brotherhood.

What is the field, and what are the raw, natural intelligences doing in the field? There is nothing material at scales of nanometers and lower, but empty space, along with a very strong field. This field is electric, electrostatic, magnetic, and gravitational, as depicted by science. Yet it is only one type of field found at these distinct states, the electromagnetic field, detected in all these separate natures of field under all circumstances, as electric, magnetic, electrostatic, and gravitational states or natures of the field. Now we can add another state capable to support life and intelligence, the interconnective state or nature of the field, carried by the field in the entire intelligent modulation of all its states, with the entire ionic form of life sitting right on top of this. This is why it is easier to study the electromagnetic type of life, because we can actually see it taking place. This is how there is life at the basic levels of the field, in the intelligent modulation of the field, which is the essence or source of life found everywhere in the field, at least here in this world. This is what we are searching for, to place it at the base of our mental model of life, modelling life from then on.

This is not exactly the base of life, because this is the case only for this world, since the field itself is at the base of this world, standing right on top of the spacetime continuum. However, the spacetime continuum of our world has its own matrix at its base, which is held entirely by our higher reality. Therefore, our essence of life standing at the base of the field here in our world actually comes from our higher reality, through our matrix, and through the field, all filled up with intelligences, many coming through the field and through the matrix, right form our higher reality, and possibly from higher realities above, all the way from the ultimate supreme main reality holding Life, Intelligence, and Interconnectivity altogether, since Life herself is the essence and the origin of everything.

Life stands on a multitude of vertical and horizontal successions of interconnected realities, filled with interconnected intelligences at all reality levels, making in this manner, creating in this manner, giving birth in this manner, or only allowing in this manner individual living beings and individual forms of life to exist, be alive, and interconnect in a specialized meaningful manner. While as you notice, from a comprehensive perspective, you do not have the same individual living beings as you see them within any reality, but you have entire lifelines of existence containing all selves and intelligences forming the living beings of all realities that it forms and intersects, while this should be in the definition of life. You are mind body and soul not randomly in three distinct realities, but you are mind body and soul placed distinctly on your own lifeline, since these three are your main selves of your lifeline intersecting all your realities on a correspondent lifeline of existence, one reality in another in another.

Our robots do not have a lifeline and a lifeline of existence, because we cannot integrate them in life, while we cannot even mimic life in order to make them alive similar to all living beings. As you study closely all living beings of Life, you notice how they live life at the confluence of two or more realities, one cognitive, the other objective or material, and the other one classconsciously interconnective, in entire lifelines and lifelines of existence. We cannot do this with our robots, and therefore we cannot make them alive. We can mimic life through first level algorithmic intelligence and through predefined interaction among themselves, yet this is not life, but normal algorithmic computation and algorithmic interaction. However, as long as our robots miniaturize and take us to the base of the ionic form of life, we are always thankful.

Furthermore, our robots do not have a third level status of intelligent beings as humans have or should have, while they do not even have a second level intuitive status as animals do. If our robot community dies, it is not exactly our fault as creators, because our robots lack living abilities in the first place. Our model of a created living being is still accurate, since living organisms and entire species always die and go extinct eventually, as it always happens due to their incompetence of coping with the environment at one time or another. You will also die for this reason, for the lack of ability to act on and react to your environment in order to save yourself from death, taking place due to some illness, shortage, dreadful plot against you, accident, ignorance, constraint, duty, intoxication, revenge, or all these combined.

For you, it is not only the physical environment or the exposure to the elements killing you, but you have to cope with society, consensual or not, since you are a conscious being, and your species had already filled up the planet, making your life highly demanding. Study life closely, to see how all living beings live life in groups, families, tribes, entourages, and communities, only to survive, subsist, develop, and prosper. Our robots still survive, temporarily, just by being able to share tasks and experience, mostly by sharing solutions of how to overcome all encountered problems.

Let us still try to climb to the second level of life, which includes the second level intuitive intelligence and the second level free interconnectivity. Let our robots that are specialized in repairing and modifying the sequence of holes throughout punch cards or punch belts to use their usual first level algorithmic thinking to modify all parameters for all robots according to the environment and according to past events, helping them in this manner to share information relatively freely, and therefore helping them interconnect comprehensively.

Because now, when temperatures drop below freezing, there is already a specific sequence of code allowing all robots to modify the viscosity coefficient of their hydraulic oil, which is very helpful. In this manner, our robots adapt to the environment not randomly, but in an algorithmic, mechanical, deductive manner, which has the advantage of being highly precise, yet it still lacks comprehensive intuition. Because deductive and inductive thinking along with repetitive thinking are algorithmic, mechanical thinking of the first level, while intuition is second level animal thinking, since it involves feelings. While intelligent conceptual comprehensive reasoning is of the third level, involving intelligent concepts linked directly to the natural laws of the universe.

The difference now is that our robots are able to generalize, improvise, and use older solutions from their database in order to solve new problems, as much as they can. If the new problem is not entirely solved, then the remaining inconvenience is treated as a new problem, while older solutions are applied as best as possible, until everything is solved entirely. The successful working algorithm is shared with every robot, and stored in their database of old working solutions.

In this manner, not all robots solve new problems similarly, and therefore these new solutions are not generated repetitively, but in a rather mechanical, deductive, semi-intuitive way. While if we had a zillion communities throughout a zillion environments, all comprising of a zillion robots, as a whole, we had perfect intuitive thinking, assuring success. This is never the case even in nature, even if cells are composed of trillions of cellular components, organisms are composed of trillions of cells, while ecosystems are composed of billions of living beings, while even this is not enough to assure always successful intuitive thinking, always successful intuitive interconnectivity, and therefore always successful intuitive life. This is why life has in this world intelligent humans, in order to assure the living success of the entire ecosystem and of the entire world at the third intelligent level, yet with all humans consensual at the first level and addicted at the zero level, it will never happen.

Let us give to our mechanical robots an extra chance, and let us wait for them to develop, as mechanically as they do. Yet we act on the randomness itself, trying to make it certain. Now we do not wait for our robots to adapt randomly anymore, but we allow them to adapt with a very high probability to all significant changes in the environment, through their own mechanical means. This is not the theory of evolution anymore as it is stated by science, but we are closer to genuine development at all levels and forms of life, as it happens everywhere.

As stated, what happens everywhere in nature is different than this simplistic patch that we apply now, since everywhere in the wider world life transcends realities one after another, realities coming in very high numbers many times simultaneously, realities that include the entire environment that life must cope with, in comprehensive, distinct living ecosystems. Life is defined not by one physical body and one intelligence held by the physical body in its own inner reality, but life is defined by a multitude of realities linked vertically, with realities creating inner realities that create inner realities that create inner realities. This type of long sequential vertical existence defines life many times in larger lifelines of existence, with the first or last inner reality tapping into the elementary raw intelligences of the field, as through extraordinary existential roots, to take its life and intelligence from there, from the source or essence of the field. What chance do our robots have to tap directly into the living intelligent field? No chance at all, unless the living intelligences of the field themselves tap into our robots, destroying them, corroding them, and taking them away from us to use their ions and molecules as they please, in a living intelligent manner, integrating them entirely in life.

We have assumed unknowingly that our robots can survive longer in an already terraformed environment, which is the case for as long as humans are around, since humans do not adapt to the environment anymore, but humans adapt and maintain the environment on their behalf, artificially terraforming it according to their needs.

Are our robots capable now to survive in an intelligent manner, and not automatically in an algorithmic manner? Yes, certainly, because, once given the ability to change each other's codes in an algorithmic deductive manner, while sharing all their new solutions, they improve, they learn in this manner from their experience, and therefore they adapt. In time, nothing stops them from finding solutions to all encountered problems, since in a terraformed environment, the number of problems and solutions is large but still finite, because humans still tend to the environment, allowing our robots to learn to cope with all current finite environmental conditions, while this is not a natural living existence, but only a terraformed one. Study humanity closely, to see how the Consensual Matrix maintains its environment finite and relatively predictable, while farmers do the same with their crops and domestic animals, helping them survive and develop, for financial purpose.

This is an artificial intelligence of the first level, by default, since the environment itself is maintained artificially at the first level. We did provide our robots in the beginning of the model with the original coding, which could have simply matched our first, second, and third level human needs, everything that we have studied coincidentally in the first book of this series, "The Human Needs."

We have just created our model of a living being in an artificial environment, and now we are more than ready to run it. I create mental models in all my books, yet I do not spend too much time running them. Now it is different, since we have much to learn about life, knowledge that is impossible to reach only by observing the life around us as science does.

Why making mental models? There are three major manners or levels of performing research. These levels of research follow the hierarchy of everything that we encounter throughout this entire book series "Human." I discovered the hierarchy of everything, and as I have designed it, it remains compatible in numbers for each intelligence and living being everywhere in all types, forms, and classes of life.

The third level applies to humans, because the human conscious intelligence is at the third intelligent level, or should be, since humans have third level intelligent potential. This is how humans have a third level awareness spanning the globe. Humans should have third level needs and development, yet not too many people identify and fulfill them. Humans should have third level intelligent conceptual reasoning, which is not used too often throughout life. Humans should have a third level intelligent society and intelligent lifestyle allowing them the fulfillment of third level intelligent human needs, but they do not have these, since the human interconnectivity is currently at the first consensual level. Humans should have a third level intelligent responsibility spanning the globe, making sure that there are no conflicts, shortages, sickness, exploitation, starvation, inequality, suffering, loss, pollution, class segregation, and discrimination in the world, while all these still take place in the world.

My hierarchy of everything applies to everything characterizing, being, or only related to life, intelligence, and interconnectivity, directly or implicitly, as status, rights, environment, conditions, habitat, interconnectivity, life, intelligence, ideology, consciousness, development, lifestyle, civilization, behavior, meaning, achievement, thinking, feelings, knowledge, learning, research, and ideas, with all numbers always correspondent.

Since humans perform research study, then these research projects can be hierarchized now on three different levels in what it concerns humans, since humans live life on the first four levels out of the eleven distinct levels determining the hierarchy of everything.

The first level research is algorithmic and empirical, based on mechanical logical thinking, as this always takes place throughout the current science, with scientists selecting previously consensual knowledge to combine it by using algorithms of logic in an inductive and deductive reasoning, forming in this manner new consensual knowledge of the first level also out of the previous knowledge, as variations on the same theme, never leaving the set of consensual knowledge that is also of the first level, since this is the agreement within the current scientific consensus. This happens in all ideologies and jurisdictions, with all theories, beliefs, and laws remaining within their own set, ideology, consensus, and jurisdiction. Since the previous knowledge had been already consensually true by agreement, then all results become consensually true just as well, by the same agreement or consensus, which in science is called scientific consensus, while in this manner, the current human knowledge never exits the Consensual Matrix.

Therefore, through a first level empiric algorithmic ideological consensual research study, you are widely accepted by the world of science, since you always remain within the consensual science, and therefore within the Consensual Matrix. All ideologies are the same, of the first level, constraining you through their distinct beliefs to remain within their own sets of beliefs, and this defines ideologies. It is actually dangerous to step outside your ideologies, and you never do. Furthermore, the entire Consensual Matrix is similarly bounded, to maintain you within its own sets of ideologies, jurisdictions, hierarchies, demands, consensual existence, and consensual knowledge, as this separates it from the matrix of life. This is what it claims, that it is always separate form life and from the real world, yet the Consensual Matrix always interferes drastically with life, this world, and with most of the wider world.

The second level research is based on intuitive thinking of any kind, as this might or might not be accepted by the mainstream of science. Intuitive thinking and intuitive research are based on intuition, which has feelings embedded in it. Therefore, every time you observe anything and it seems right to you because you feel it in your guts that it is right, then it must be true according to your own second level intuitive judgment and feelings, and you state it out loud and clearly for the entire world to hear it, that you already feel it in your guts and how good it is to feel it in your guts, since you must always be right according to your feelings at the second intuitive animal level. This is the alternative science, considered apart from the mainstream science, yet still in the Consensual Matrix.

Intuitive thinking is the lowest form of free thinking, standing apart from the multitude of ideologies, hierarchies, and jurisdictions. We find this second level thinking throughout the animal kingdom, since this is the basic animal thinking. While we also find it in the alternative science, slightly, because the alternative science is also consensual, more ideological, and less algorithmic, making it rather superficial, or even ignorant. While being controlled by the same Consensual Matrix, keeping it at the same first consensual level, but not at the second intuitive level as the alternative science claims.

Humans are different, since humans have the chance to use intuitive thinking on conceptual knowledge in order to obtain new conceptual knowledge. Human reasoning and animal intuition remain incompatible, but no one ever cares. Currently, people still use intuitive thinking on intelligent conceptual knowledge while decaying everything to the second intuitive level, because science and psychology cannot model any of these. This is a major error of reasoning, generating all problems, since it constrains humans to use the intelligent human mind in any manner, even intuitively and consensually, without knowing too much about the intelligent human mind. Which is dangerous, since humans end up enslaved, addicted, sick, disabled, or they end up living life as all animals do, and you can see it everywhere, while coincidentally, making the Consensual Matrix possible here in this world.

Why is the second level animal intuitive knowledge not compatible with the third level intelligent conceptual knowledge and reasoning? Because humans should use their third level intelligent reasoning for intelligent conceptual knowledge, not their second level intuitive thinking. You can use intuitive thinking throughout the fulfillment of all second level physiological needs, as eating, security, recovery, and social needs, since you have within yourself innate solutions for fulfilling many of these needs, along with your own successful rational solutions, and this is how you have your feelings in your guts with the fulfillment of all your second level physiological intuitive needs and meanings.

Let us study dogs shortly. Dogs are carnivores because their intestine is too short and cannot digest anything else but meat, or this is what everybody recites stereotypically throughout school. You can even feel it in your guts that it is true, at the second intuitive animal level. This is an empirical or intuitive statement, making a random analogy between two characteristics describing dogs, as being carnivore, and as having a short intestine. Yet these characteristics are backwards on their own line of causality and therefore on your own line of reasoning matching it. Because in the first place, all dogs are carnivores, all carnivores have a specific type of digestive system adapted for meant, while this happens to be shorter than the digestive system specific to herbivores and omnivores, since these eat plant-based food and need a longer and larger digestive process to break down the specific cellular walls that the cells of many plants have.

This is the concept associated to the shorter intestine that dogs have, while this entire line of reasoning is conceptual, at the third intelligent human level. Do you see how you have to involve a combination of all levels of thinking, only to be able to model comprehensively the dog intestine? Otherwise, you end up with an empiric, intuitive, and even ideological, stereotypical, consensual research, showing that dogs eat meat because their intestine is unfortunately too short.

Study the current mainstream and alternative science closely, to find only first level and second level research, with alternative science specialized in intuitive, empirical research, and you see it everywhere on Discovery Channel, History Channel, and Animal Planet. It is worse than our little robots, because all scientific research should be comprehensive and intelligent in nature, at the third intelligent level. Yet you always find lower level research taking place everywhere throughout science, because science does not even perform research as it claims. The Brotherhood and therefore the Consensual Matrix demands the research itself, the manner in which to perform research, and the specific results to obtain during research, while the current consensual science obeys. While if you are a more capable scientist, always persisting to invent the most efficient sausage machine in order to make the world a better place while eradicating famine itself, you lose your job. Yet currently, you are never employed as a scientist if you are not at least in the Lower Brotherhood, while this is worse than in our model with robots. How does humanity as a whole manage to withstand the environment? The Consensual Matrix tends to humanity just as farmers tend to their crops and animals, making sure that these are never harmed by the elements, while exploiting them minutely.

If dogs are not too relevant, let us study polar bears. Currently, mainstream documentaries state how polar bears drown because there is no more ice to live on, and this is how they go extinct, blaming global warming. This is the scientific propaganda, and you encounter it often. One specific scientist stated in a documentary, while observing one specific polar bear in the Arctic, which was barely clinging on a little pad of ice, that polar bears fall in the water and drown. You find these documentaries everywhere in the media, all pretty in colors but based on empirical intuitive ideological consensual research.

Later on, the polar bear is shown swimming while catching his fish and seals, while the scientist states that water is very cold there since it is in the Arctic, and it must be terrible for the poor polar bears to swim. Yet the water must be around zero degrees Celsius, since saltwater freezes slightly below zero Celsius. You can find water this cold during winter everywhere in Europe, North America, and throughout all temperate regions of the world, not only in the Arctic.

Our polar bear comes out on ice now, and starts rolling and dragging itself in the snow everywhere, in a rather interesting, playful manner, making us smile. Was he not cold in the water? No, the scientist states, because of the effort that he involved while swimming, being such a large mammal. The bear is very hot, and it must cool down on ice. This is what the scientist states, since this is how all polar bears cool down after a big swim, rolling directly in the cold snow, because luckily, they still have snow in the Arctic.

Notice how everything in this research is empiric in nature, based on direct observations and on speculations, while the scientist manages to use the specific feelings embedded in all presented information in order to coordinate people's intuitive learning towards accepting his erroneous statements as being accurate, since people can feel them with their guts that they are accurate, and therefore they must be accurate. This is called indoctrination, and it takes place in all ideologies.

What is wrong in this documentary? Everything, since everything contradicts previous statements, as these are explained empirically, as best as the scientist could do intuitively and consensually. Because it is a fact that you can dry fur and cloth instantly by rubbing them directly in very cold snow, since very cold snow and very cold ice cause the liquid water of the fur or cloth to freeze instantly, shatter, and detach from the fur and cloth very quickly, ice-drying them in this manner almost instantly. While by dragging yourself in the icy snow, you brush off the newly formed ice, ending up with a dry and clean fur or cloth, in a matter of seconds. Try it, but it has to be very cold snow or ice, or you end up with mushy wet snow all over you.

The third level of research is the comprehensive intelligent research, based on comprehensive intelligent mental models, comprehensive intelligent abstract learning, comprehensive intelligent study of all lines of causality involved, comprehensive intelligent digestion, continuous intelligent elaboration, and intelligent mental modelling.

Mental models of higher levels cannot be stated directly in books or documentaries, because books of nonfiction cannot offer you more than enumeration of data, along with first level algorithmic lines of thinking, as these can match only first level knowledge. While life can have concepts of up to the tenth supreme level associated with it, and cannot fit in books. The understanding of the human reasoning can be done through third level intelligent concepts, and can fit into third level intelligent mental models, while these form directly in your intelligent conscious mind.

We cannot place third level intelligent mental models into books, since books can carry only first level knowledge, first level concepts, first level communication, and first level ideological or algorithmic lines of thinking, through normal book limitations. Textbooks are worse, because they are always at the first algorithmic and or ideological level. However, if you know how to write your books, you can form entire third level intelligent mental models directly in the minds of your readers. It is the same throughout education, because good teachers are always capable to help students conceive, form, and maintain third level intelligent conceptions and intelligent mental models in their conscious mind.

The second level mental models are intuitive, very similar with your daydreams, novels, and romances. Which are very similar with all books of fiction and with all movies that you can watch, all being at the second physiological intuitive level, based on your intuitive feelings and intuitive thinking. While all second level intuitive mental models are formed not inside books, but in your mind, when you read the books, or when you watch the movies, because you are the one attaching your own sets of needs and feelings for each object, subject, scenery, plot, and circumstance from the book or movie. The mental model is in your mind, but not in the book.

It is similar with intelligent mental models of the third level, since these are in your mind, as you form them while reasoning continuously at the third intelligent human level throughout the book and in parallel with the book.

Right now, you are forming your own mental model of life, in parallel with this book. As you can tell, it is very complex, because Life is of the tenth supreme level, while I model life at the third intelligent human level, since this is our human cognitive limitation. While I have to place all these words here in the book at the first algorithmic level, which is the limitation of this specific written form of communication, books.

If I chose videos instead to learn about life, I had to recite this entire book for hours in front of the camera, with similar results. Yet if I used videos and images throughout the video, along with actors and computer animations, then yes, it would have helped you more throughout your intelligent mental modelling. Yet since my books are already numerous, and since it takes my entire life to research them, model them, elaborate them, write them, edit them, and reedit them repeatedly, then books are my only option and your only chance, and this is what you read. Yet if you have made it so far in the book and book series "Human," if you can follow my books with your own lines of reasoning at the third intelligent conceptual level, and if you are rewarded continuously with love and happiness for your success in learning and development at the third intelligent human level, then yes, it works, developing you and the entire intelligent human environment.

It is the same with research, because you have to match the level of each concept that you research with the level of your research. When you research how you witness and follow your entire zero level addicted experience, you can do so at any research level, since your specific firsthand experience is of the zero level, addressing mostly people of the zero level who live their life addicted. It is the same with porn, the basic sex for pleasure, as long as it is not meant for reproduction at the second intuitive, physiological level, but only for zero level addicted pleasure. This is a perfect example of zero level research: explicit data recorded into videos, and then further zero level studying done by everyone interested in zero level topics, the entire world, while the Internet if full of these, stating precisely how oppressed humanity is from reproduction.

You can tell right now how you cannot watch zero level porn with your third level intelligent cortex, since it tarnishes the entire experience. Similarly, you cannot read second level romance and novels in general with your third level intelligent cortex, because again, you cannot experience fully the entire soap opera from the books with all the circus involved, since you cannot feel it in your guts and everywhere else in your body, because you have to match your cognition with everything that you study or experience.

Furthermore, this is how you cannot use your own third level intelligent conceptual reasoning throughout first level consensual lodges, mobs, courts, ideologies, and jurisdictions, since it insults your intelligence, while it challenges your need for justice and equality in the world. If you still happen to have a third level intelligence and responsibility throughout lodges, ideologies, mobs, and jurisdictions.

First level research is empirical, consensual, algorithmic, legal, and ideological, and can match any book, documentary, and textbook. The current science is entirely of the first level, making all textbooks available in large numbers, at the first consensual ideological level. It is very easy to conduct first level consensual empiric research, because you simply observe your phenomenon, you record it in any manner, and then you can give any explanation, as long as it remains consistent with your empiric observation, which is only the last effect of that specific line of causality observed empirically. With you free to invent any cause you can desire matching that last effect that you observe or invent, as in the case with the big bang theory.

This world is filled with lines and lifelines of causality, as you should focus your research on them entirely, but not only on the last effect, whatever manifests right then in the world. Yet as you study entire lines and lifelines of causality, you have to match them with your own lines of algorithmic thinking and intelligent reasoning, and with the specific natural laws of the universe determining them.

As a reference, it takes an entire cortex to make possible the third level conceptual intelligent reasoning, distinguishing the third level intelligence from the basic intuitive thinking of the second level. The third level intelligent reasoning takes place in the third brain, which is the human cortex, while the second level intuitive thinking takes place in the second brain, which is the midbrain, or the reptilian brain. This is why you must perform third level intelligent research, because living human beings are of the third intelligent level by nature. More precisely, if you have a cortex, then use the cortex at the normal third intelligent level.

The problem is that you cannot record third level intelligent research results into books, scientific papers, thesis, and documentaries, but only into comprehensive third level intelligent mental models found directly in the cortex, because nothing in the world can form, make, create, or conceive third level intelligent knowledge, but the human cortex. Yet since intelligent knowledge and intelligent reasoning are not in the real, objective world, but only in the conscious mind, you cannot patent them, you cannot own them, and more importantly, you cannot constrain people to abide by them. This is why you cannot use higher level intelligent mental models in the Consensual Matrix, but only precise enumerations of knowledge, beliefs, agreements, laws, rules, and regulations, as these are of the first algorithmic and consensual level.

While these constrain third level intelligent human beings right into the first level consensual Cinderella shoe, which is dreadful. While when it happens for generations, for centuries, for millennia, and for entire dark ages, it is not coincidental or unfortunate anymore, but it is business as usual, normal tyranny and normal profit.

Notice how, by using only two words: Cinderella shoe, you form directly a second level intuitive mental model in your mind. One size fits all, which is another definition of the current consensual human society, shredding continuously your intelligent human mind in order to fit the consensual Cinderella shoe.

With the last sentences from above making a better second level intuitive mental model in your mind, since they are capable to offer more information than the amount of information that their words state individually. This is how mental models are formed, in parallel with everything that you learn and experience, as you augment continuously your conscious intelligent inner replica of the world spanning the cortex.

In the Antiquity, in Europe, people researched and wrote in

Latin, leaving behind countless of books written at higher cognitive level, mostly at the second intuitive level, capable to offer significantly more than what their words were capable to state individually. Yet you had to know the specific events, habits, values, and circumstances of those times, in order to be able to form your higher level mental models when you read them.

How accurate are our intelligent mental models? The more our models resemble reality at regular settings, the more consistent they are, and the more we can count on them when we run them later on, through any assumption and at any setting, in order to find out anything of interest, to find out about artificial learning, about the origin of life, about the development of life, about life after humans, about other forms of life, about life after death, and about life at absolute supreme levels, since once the model of life is formed in your mind and it is accurate, it can offer everything, while this is normal third level conceptual intelligent reasoning.

Before we run our model of a created living being, we have to set all details in place. Since once we start the model, we cannot intervene, or we risk influencing it, compromising the results.

We decide to start with a few dozen robots both identical and specialized, placed in a vacant, neutral environment, with all parameters set to regular, or normal. We expect our robots to be able to cope with the environment, mostly by using our specifically provided solutions, which they store in their database. Thousands of solutions for any problem that they can ever encounter, solutions allowing them to learn from each new event if these manifest, and to add these new solutions to their database. Additionally, our robots must avoid unfortunate events, by identifying them, predicting them, and then by preparing well in advance to confront them if necessary, even by modifying themselves in any manner, in an entire developmental process that we call adaptation. Our model is going to challenge our robots in a random manner, through a multitude of good and bad environmental events and conditions, each one with the appropriate probability of occurrence matching our environment here on Earth.

Be prepared, since many times, after a tedious effort, your mental model can give you wrong results, trivial results, or no results at all, nothing but a flat straight line on all diagrams, which might depend on many circumstances. In general, the first dozen iterations are used to adjust the model, yet some models tend to run relatively well from the beginning.

Let us now start our intelligent mental model of a created living being. We do not expect it to lead us directly to the organic form of life or even to living chickens and living human beings, we do not expect it to find a brand new type of life, yet if our mental model of a created living being takes us to the base of the ionic form of life while also helping us learn more about meaningful specialized interconnectivity, about creating entire new forms of life right on top of the old ones, and about the intelligent living oscillations taking place right at the base of life in the raw electromagnetic field, then we are very thankful.

We let our robots free in the desert, and they start immediately to fulfill all their needs, one after another. They spend more time in the sun in order to charge their batteries, they start looking for adequate shelter, and they search for resources to build their parts.

Days pass, and our robots manage to fulfill their needs, mostly. They survive the strong wind, and now they are cleaning each other of sand. Weeks later, our robots are able to cope with all casual weather, and suffer very little damage, which is very good.

Note how easy it is to study a model and not the real thing, since models can be stopped and fast-forwarded at wish, while they can last as long as needed.

Years later, our robots have learned everything about their environment. They have built themselves a solid habitat and they can predict now most of the unfortunate environmental events, taking shelter before they occur. Our model is a success, since our robots manage not only to survive, but also to learn, to add to their database of solutions, and to share among themselves all experiences. They have also replicated themselves, once, when an unfortunate falling bolder destroyed a cleaning robot.

What concerns us is the very low amount of learning that our robots undergo. We were expecting them to develop into a genuine artificial new species of Earth, but our robots choose to avoid every event, unfortunate or not, and just take it easy, to save energy and to preserve themselves as much as possible.

Why is this concerning? Because there are significant disasters coming randomly their way, many capable to destroy the entire community. Our robots have no access to our database of environmental events challenging them randomly, they had never encountered major disasters before, and from the way they behave, they stand no chance to survive outstanding events still to come, since their level of learning had already dropped to zero, after mastering their habitat throughout all possible mild events.

We keep running the model, and one century later, nothing changes. Our robots only replicate themselves occasionally, whenever individuals are lost, and then they continue living in perfect equilibrium with the environment. Hundreds of years later, a minor earthquake causes an avalanche of boulders to fall down the mountain destroying most of our robots, including all robots specialized in building parts. The robot assemblers survived, yet without parts, they cannot replicate new robots, and when these perish, the entire community is lost.

This is what life strives to achieve, homeostasis, the perfect harmonious equilibrium with the environment, when you have all the needed resources and when you are able to learn and to predict all events. However, it seems that this never happens, since you are never able to predict dreadful events, including the most destructive ones, which tend to happen less often. What Life does in these circumstances, it makes everything and everybody reproduce, divide, or replicate rapidly, to cover the entire environment. When local cataclysms occur, only parts of the entire species are lost, while the species itself still survives.

Yet when you have a multitude of species around, overpopulation becomes an issue, and therefore species can spread out only as far as their own niches last, and only as far as the other parallel species allow them to expand. The problem is not exactly that species are incapable to adapt to the environment and therefore they go extinct, since many times, their niche is altered or it is not available anymore in the environment, and they either develop, or they perish. Because each change in the environment and each loss in a niche are opportunities for newer intelligences to take over brand-new specializations within the cell, organism, pack, society, species, or ecosystem, and therefore it is an opportunity to exist. In our case, if our robots were genuinely alive, then there were already newer intelligences emerging to take over the newly available existential niche, as the part builders. Yet without living natural intelligences, nobody was able to replace the lost type of robots, the part builders, and the entire community of robots perished. While we cannot help our robots with natural intuition either, since the current technology cannot offer it, which means that our little robots are still not alive yet. Therefore, we keep on modeling, to see what happens in the next iteration.

We do not only play with robots, but we use the current technology at its best, since this is how we are able to use actual assembling robots in our model, because these are currently used throughout the assembly lines of many factories in the world. You can find these assembling robotic machines in all modern plants, mostly in car factories, and we can use them in our model, while their punch belt computers should be capable to record all their experiences and solutions, old and new. This is not exactly the case in real life, but it is the best that we can do.

Let us now conclude that our robots were built of a material that was not strong enough, and this is why the boulders had crushed them. Furthermore, we were very close to surviving, because if only one assembler robot had survived, now our community of robots was still alive, learning how to prepare themselves in order to cope with the following landslides, probably by strengthening themselves or by learning to run faster, getting out of the way of the incoming boulders systematically.

We increase the coefficient of material resistance for all robots and we rerun the model, hoping that they survive this time, through stronger bodies.

Our new very strong robots show a good start, and they tend to spend less time on building a strong habitat and additional robot parts, since they are strong enough now to confront the environment out in the open. Their frames are stronger, yet most of their components are still weak and are damaged easily. In only decades, our robots use the transparent plastic material from their lenses to build plastic punch cards, which are stronger and significantly smaller, recoding significantly more. Furthermore, instead of having a lever checking each hole from each card, now our robots use the technology from their video camera to read each hole optically, not mechanically. Due to their mechanical intuition, our robots do not use the best materials and the best technology integrated in themselves in order to build this rudimentary data storage CD from scratch, but they only manage to build a modified, adapted version of an entire video camera, and they use it to read their sequences of punched holes from the plastic cards.

Life does the same throughout the multitude of her species while these develop, since she has to recycle entire structures from a living being that are either physical or cognitive, while she does not use only the needed individual parts, which seemed more appropriate. As we learn more about Life, we notice how everything within her species and living beings is placed one on top of another, resembling exactly to homemade appliances.

Everything relates with the specific intelligences living within these systems of the living being, because these live in the entire system, and not only within isolated parts. This is how Life does not recycle bodily parts to use them to adapt to future unfavorable conditions of the environment as seen in the physical world, but Life actually recycles or re-specializes entire systems of intelligences to solve the new problems, exactly as these are found in their physical bodies. These highly successful intelligences capable to solve new environmental problems are kept as prototype intelligences within the chromosomes of the DNA.

We can observe this developmental tendency in Life herself, since species do not use specific information present in previous adaptations, but they adapt the entire system to new tasks and circumstances, making the entire organism appear rather primitive in ingenuity, while the information is there continuously to offer better developmental ways. This is how the brain itself is actually packed up skin, as it is not constructed from scratch using structural information from skin, but using the entire skin itself, folded up upon itself, and most importantly, using the conscious intelligences packed up in the skin, now becoming the conscious intelligences of the entire organism, you the conscious intelligence.

Intelligences are not exactly held within the physical body, as computer hardware holds computer software, but whatever exists subjectively within the body, is seen objectively as the physical body. The intelligences themselves are the physical body, as seen from an outer, objective perspective. More precisely, the inner realities holding all intelligences are seen from the outer, objective perspectives as physical bodies. When you have physical bodes and intelligences as one, you have life, while this should be in the definition of life.

This does not only improve the strength of their mechanically recording device, but it makes it both smaller and more capable to record more. Consequently, one thousand years later, our robots are about a tenth of their original size and a hundred times more numerous, yet they had also stopped learning and improving centuries ago, as in all the previous runs of the model. Now we just wait for the following major cataclysm, to see if they survive it or not, by being stronger and more numerous this time. Why exactly did our robots stop learning and developing?

Our robots survive now from one major cataclysm to another, depending on how drastic these cataclysms are. This is the case with humanity throughout the ages, as we learn it from very old records and from religion and spirituality, since humanity tends to die out at the end of each age of Earth, reborning afterwards from its very few scattered cultures.

Are our robots capable to survive the next cataclysm? Are humans capable to survive the next cataclysm? Our robots have already survived the past one, and they have learned greatly from it, remembering now all sets of solutions necessary to keep them going, or to get their community going in the future after the following cataclysm.

Are humans ready for the next cataclysm? No, not at all. Humanity remains diverted from one day to another, and has no chance to survive even milder cataclysms, as basic polar shifts. When electricity goes out in many civilized nations for more than a week or a month, people are so undeveloped, that they kill each other for the last cracker in the grocery store, and that was the last cataclysm. Our robots are spared of the continuous social sabotage that humans inflict on themselves, and therefore our robots have by far more chance to develop, adapt, and survive. Job well done.

Organic life develops and adapts similarly to our robots, while biology is incapable to explain how and why. Study organisms closely, to see that they never invent anything from scratch, the way humans write books or invent technology, but everything new is adapted, everything is an old invention used for relatively similar tasks but under different circumstances. I am not using randomly this example with the video camera or with robots, since nothing in my books is random, trivial, or already studied, otherwise it would be unnecessary for you to know, wasting my time. Through this example, I only stress on why organic life develops as it does, through which procedures, and through what type of thinking.

Organic life is and behaves similarly to our robots because

it uses the same model, it is placed under similar presumptions and circumstances, it is targeted by similar environmental conditions, and surprisingly, it uses the same initial type of intelligences, yet life is infinitely vaster than our robots. You can still notice similar traces in behavior and development with our model, even though our model is a different type of life, of a very low developmental level. As stated, your brain is made from regular skin, folded on itself. It is a solution found long time ago, when the organism needed an organ to guide and synchronize all cells of the organism according to the outside environment, built mostly around its sensors of perception. The skin itself is made throughout consistent development, yet initially, it was cellular membrane. We still notice the same structural traces of cellular membrane components in the skin and therefore in the brain, because these are the conscious intelligences specialized in the continuous interaction with the outside world, since everything relates with the specific specialized intelligences inhabiting all these recycled bodily structures.

In the case of the cell, its own conscious cellular intelligence specialized in the interaction of the entire cell with its outside environment lives in the cellular membrane itself, now recycled and re-specialized in the skin of the organ or organism. This same cellular conscious intelligence has to monitor and coordinate the interaction of the entire organ or organism with its outside environment, while it fulfils its needs. Afterwards, when organisms develop brains, they actually divide all cells in time, to make a large quantity of skin, with all its conscious intelligences still living within, to make the brain. Homemade appliances.

Do you see how it remains irrelevant how the physical body actually looks or ends up looking after the required adaptation or improvement? Because it matters only how the intelligences present within that improved structure of the living being are capable to fulfill the new set of tasks while coping with the new environmental conditions.

How exactly do these newly re-specialized intelligences

actually know how to cope with the incoming drastic environmental conditions? This is why they are used in their new adapted specialization in the first place, because they are the only ones that the cell, the organism, or the entire species has, with whatever they know being the best ever known. They are the best hope, and now only time can tell how successful they are. With additional similar development and adaptations taking place continuously, ending up to have exactly the living beings and intelligences that you see everywhere.

Are they capable to perform the new tasks, relevant to all changes in the environment? Yes, so far. Because if not, they have to resume only to the specific niche that they are capable to reach, which might not be enough. However, there are always species capable to reach all niches of Earth, while only under extreme circumstances, entire species go extinct, to make room for those who are even more capable.

The entire process of development is not exactly a race or competition between species as these cling to the last niche available in the environment only to survive, but since all these specialized intelligences old and new are more faithful to Life than to their own cells, organisms, and species, it is Life always prospering from all new successful specialized intelligences, through one specific species or another, enhancing her own living abilities continuously, and this is Life.

Are our robots similarly capable? This is what we are trying to find out, and this is why we are running the model of our robots throughout dreadful cataclysms, only to find out how they are capable to adapt. From what we can assume right now, intelligences are always eager to re-specialize with any new change in the environment, good and bad, since these always bring new niche elements in the niche of the entire species. While if you only know how to make use of the new niche element as an intelligence, you find yourself a specialization within the entire cognitive system and therefore within the entire organism. Environmental changes might be dreadful, yet they are always new opportunities for new specialized intelligences to prove themselves useful, and therefore to exist from then on.

From the cognitive perspective of all intelligences, all changes in the environment are not exactly unfortunate, but they are opportunities, offering new specializations to new specialized intelligences, if they only know how to take advantage and perform the new task. Was the oxygen event a good or dreadful event? Oxygen is toxic and very corrosive, ruining all anaerobic life. However, for all aerobic intelligences knowing well how to cope with oxygen itself, it is certainly an opportunity, as it is for all human beings and for all current plants and animals of Earth, because everybody is aerobic and loves oxygen.

More precisely, these new opportunities are entire new niches offering everybody life, if they only know how to reach them and use them. All these new intelligences are based directly on the raw intelligences of the field, and therefore they are alive themselves, part of Life, which is the case with all living beings, of all forms of life.

There is nothing random or malfunctioning making all new specialized intelligences possible in the living world, but only continuous successful interconnectivity, continuous achievement, and continuous success. Since all related specialized intelligences help the new specialized intelligences in every manner, through the same needs and feelings that you know well.

How can we tap into the raw intelligences of the field ourselves? You have to be alive in order to do so. While as a living being, everything new that you learn involves a new task and a new intelligence to tend to it, new intelligence that, through all its cognitive abilities, is a system of intelligences in itself, composed of a multitude of inner inner raw intelligences of the field. These are natural intelligences of the field coping with changes in the environment, caused by similar raw intelligences of the field, in an apparent cognitive conflict of interests.

How exactly do these new, re-specialized intelligences know exactly how to do everything, and why would they ever get involved in this tedious work? Because everything that intelligences do, they do to fulfill needs and meanings. It is the same with you, and it is the same with all living beings of all levels and of all forms of life, all fulfilling needs and meanings throughout life, and nothing more.

The difference between needs and meanings is that while needs refer only to the lifetime behavior and activity of the individual living being, meanings in life go outwards, to become a source for the others to fulfill their needs, addressing the entire community, species, society, and environment, since meanings are normal needs coming from higher classes of life and higher forms of life.

Just as you are eager to save yourself, your family, your community members, or the entire nation and world in any circumstance, through your inner natural needs and meanings, all living beings and intelligences do the same, as they use their entire reasoning, pertinence, abilities, awareness, and knowledge to solve all security problems, while these needs and meanings come straight from Life or from the Field, through inner intelligences through the field and through the matrix, through the higher and higher realities, all the way up to Life herself, since all life is interconnected, it is Life herself.

As we start noticing, there is more to consider, since the environment itself that changes continuously is also part of Life, if it is of a different form of life. It is also organic, as it is the case in society, creating mostly unfavorable consensual conditions. Only Life knows what happens next, while developing all its intelligences accordingly, at all levels. While from the perspective of all intelligences, it is always only a matter of being able to fulfill all their needs and meanings within their environment or not. While when you are very good at fulfilling one specific need, or if you are the only intelligence capable to perform this specialized fulfillment, then you become specialized in the fulfillment of that specific need, it becomes a meaning for you and a source for your entire form of life, for the entire organism or the entire society depending who you are, and now you are priceless, you are who you are, and you are always meaningful and fulfilling. Job well done.

It is not as modeled by the current science, with one unfavorable environmental condition killing the unfit and spearing the rest because they can cope, but it is always about life integrating perfectly within life, even harmoniously, since the environment itself is alive, always integrating more life within in a harmonious, living, meaningful manner. While with every niche gap ever created, occurring, altered, or left unoccupied, you have a zillion intelligences eager to find ways to occupy it, only for them to have a specialization, a meaning in life and in the world, a distinction, a way of living, a contribution to Life, a part in life, a part in this world, and therefore a part in existence, in order to have its own life and existence, to exist, to be alive, to survive, to subsist, and to develop, because all these are distinct main needs for all living beings and intelligences.

The new raw intelligences of the field appear and adapt to all new niche elements manifesting with each change in the environment. Yet it seems that these are not entirely new intelligences, but they are old intelligences, re-specialized, recycled, developing themselves to newer tasks.

There is more to consider, since intelligences help intelligences to develop, according to their common needs. When intelligences develop, they receive newer abilities coming from other intelligences and from other systems of intelligences, many times even changing them entirely, while these are the new specialized intelligences. More precisely, an entire consistent interconnectivity makes possible all newly specialized intelligences, not only their own effort to develop and to adapt to all newly occurred tasks and specializations.

Everything relates with the needs and meanings that all living beings and intelligences have, including the specific need to find your own specialized meaning in life and in the world. It is the same everywhere, in all societies and communities, since intelligences live life normally, as normal, objective living beings from their own perspective wherever they are, at inner and outer levels. While all intelligences and all living beings have the need or meaning to gain their own identity wherever they are, to make a name for themselves as it is called, to become something important, to do something in life, to self-realize themselves. Because when you do not achieve any of the above, you cannot achieve the need and meaning of your life, you are punished dreadfully, you feel incomplete and unfulfilled, and this is a terrible feeling. People take drugs throughout society only to overcome the feeling of incompletion, dropping directly to the zero addicted level while making everything worse, and that is not life. While when the wild animals become meaningless and unfulfilling, they stand alone since no one wants to be around them, and they soon die. It is so terrible, that all meaningless living beings offer themselves to their predators as easy food, and this is how they die. Yet insects turn themselves upside down deliberately, they cannot move anymore, and they die. While if they prefer, they offer themselves as easy food to all birds, and this is how they die.

Why do insects do everything that they do throughout life? Why do our robots do everything that they do throughout each iteration of our mental model of a created living being? Why do humans do everything that they do throughout life? Humans fulfill needs and meanings through their punishment-reward cognitive mechanism, being punished and rewarded according to their success, through the multitude of their feelings. Other intelligences punish and reward them, since other intelligences send them these needs and meanings, throughout the continuous interconnectivity within all cognitive systems, and throughout the interconnectivity taking place within each intelligence, since all intelligences are systems of intelligences themselves, with one intelligence for each one of their cognitive ability.

All developmental needs relate to specific feelings of punishment or reward, and therefore all re-specialization for newer tasks takes place through punishment and rewards, through feelings, or through needs, as part of the continuous interconnectivity within the cognitive system and within the entire higher form of life, as the entire organism, family, genetic line, community, species, civilization, ecosystem, reality, and Life. Since this is why all intelligences are more faithful to Life, because they are her intelligences first.

Development or adaptation are only a continuous change taking place within Life, just the way the field itself shifts or changes itself from one state of the field to another within itself at its own raw roots. Again, intelligence dissipates into life while studied closely, all dissipating into interconnectivity and into the raw field, then back into life, as a continuous shift or change from one state of life to another, from one mode of life to another, and from one form of life to another.

Intelligences specialize continuously, and therefore they render the entire organism, species, or society fit and adapted to the environment through their own natural needs and meanings, which come directly from life. When we study Life closely to see how everything happens, our answers relate back to our intelligences, field, and interconnectivity in the field and in life.

What is going on? How exactly do intelligences, individual living beings, and entire species and forms of life know how to adapt to the environment? Firstly, individuals and species do not adapt to a dead environment, since all environments are alive, consisting of all forms of life in the ecosystem. In your case, while living your life, you have to interconnect, cope with, and adapt only to other human beings within the human society, and nothing else. While many times, you do not even have to cope with other human beings, but only with other consensual corporations within the multitude of jurisdictions and ideologies of the Consensual Matrix, since the Consensual Matrix makes life very simple currently, and calls it consensual existence.

Yet it is always a matter of life interconnecting with life, striving to take place in the most effective manner, and not necessarily in the most harmonious and cooperative manner. Yet it always depends on the specific mode of life in which you are. Therefore, you are always in the scenario of the puddle matching the hole in the road, and this is how you find your meaning and place in life and in the world, since it is a highly complex process. Yet you always have a choice, if you do not live at the first level in servitude, since this is what characterizes all natural life, choice, determining directly all lifelines of causality here and in the entire wider world. While it takes all mental models of all books of this series "Human," to model and understand life and its cognition, meaning, behavior, development, fulfillment, and achievement.

For example, when life has to match a living environment full of forms of life, you have to consider that all life found on all sides of the environmental interaction takes place in various realities inner and higher simultaneously, while the specific reasoning that you attempt to employ is compatible only with this world, because you attempt to use only accurate facts and lines of causality that are accurate only here in this world. Lines of causality are different within higher realities, because our spacetime continuum and natural laws span only our world, since time, sequences of time, and entire lines of causality manifesting in time are different in other realities, mostly within our higher reality.

Therefore, you are not in the specific line of causality where the environment changes, resulting in species having to develop, resulting in other unfit species having to die, resulting in the best of the best making it alive, resulting in your question of how exactly did they know to adapt in the first place, even centuries and ages in advance. Living beings develop through themselves as they are within other forms of life and realities, and through the multitude of intelligences of their own cognitive system, along with the higher selves that they are throughout higher realities, everybody having the same goal, staying alive and staying fulfilling.

Because as stated, life is lived throughout a multitude of vertical and horizontal realities, while you must remain interconnected continuously at all levels of interconnectivity, in all forms of life, and in all realities. Because all realities have a cognitive nature, as this is still the case with our world. Which means that life is placed under these specific conditions not randomly, but with the purpose of constraining living beings and intelligences to find successful solutions.

Everything is a higher mental model that takes place wherever you are, even in this world, while nothing is random, as you always do everything on behalf of a higher reality and on behalf of higher beings that do everything on behalf of their higher beings, up to Life herself. Because knowingly or not and willingly or not, you are always more faithful to Life than to your own kind when you live your life at the intelligent human level, and not addicted or consensually on lower developmental levels. It is the same with your intelligences living their lives on your behalf, while you call it simple thinking, and while you are determined to ignore it mostly, as you are caught in more relevant daily activities, as watching the media, drinking, or going to work all day. This is what your higher beings receive throughout your life: media, poisonous food, indoctrination, vices, and addictions. Yet since most of what goes on down here is requested directly from up there, what exactly is going on up there, if all they want is tyranny and drugs? Can it actually be worse up there than down here? Yes, what exactly is going on?

The Consensual Matrix, this is what goes on, up there, down here, and everywhere else, throughout most of the wider world. Because we could not exactly invent the Consensual Matrix down here on Earth on our own, but the higher beings or souls bring it down here with them every time they come down, not to get in trouble up there. Vices, tyranny, and addictions follow the Consensual Matrix closely everywhere throughout the wider world, since only vices, drugs, and tyranny motivate all life within the Consensual Matrix to stay in the Consensual Matrix, and now this is life. On drugs, under tyrants, and in violence, and what a thrill, because these are the movies, life, and videogames that you seek the most throughout all your realities, and what a thrill.

This is why Life uses entire old systems to adapt species to new changes in the environment, and not only the information that it can use to build these systems from scratch. These old bodily systems used now in different parts of the body for new tasks and circumstances already contain the specific intelligences and systems of intelligences performing similar tasks before, allowed now to perform old tasks in new places and under new circumstances. Your cellular membrane contains the conscious intelligence of the cell, used by the cell to interact with its outside world. When cells form tissue, organs, and organisms, they do not simply build new similar skin to wrap around groups of cells, organs, and organisms, but they carry the old cellular membrane to new places, forming there skin as thick as necessary.

They do so not by transporting this skin physically, but by bringing to life zillions of these new specialized intelligences there in the new place of the organism, while forming actual skin or cellular membrane as these conscious specialized intelligences look from a physical perspective, skin. This is how intelligences look from a physical objective perspective: skin, tissue, horns, or bones, whatever they are at their origin. While this specific skin already contains the same old specialized intelligences, ready now to perform similar new tasks, only in a different place, and for a different meaning. It is the same when brains are formed, because intelligences are carried with the objective skin holding them, to form newer brains, as the new cortex.

It is not the body that counts, as in a mechanical manner, but the intelligences within the physical body count, in a cognitive manner. Because it is more important what these intelligences can do for the entire organism, and this is why they are being transported now wholly from one part of the organism to another throughout the comprehensive development of the entire species, done so in order to make the organism capable to make use of the cognitive abilities of the specific intelligences residing there, now transported to new places where they can perform relatively similar tasks, but for new meanings.

Organisms are more as communities or civilizations themselves, while individual intelligences and individual systems of intelligences inhabiting the organism are what counts in life, and not exactly the physical body. Therefore, if you the conscious intelligence happen to live now in a small group of neurons within your prefrontal cortex, know that you used to be a conscious cellular intelligence only several billions years ago, responsible then with the interaction of the entire cell with its outside world, while now you are responsible with the interaction of the entire organism with the outside world, while fulfilling exactly your old needs, and while performing exactly your old tasks, as finding food and taking it in, or seeking shelter and safety. You have been doing everything successfully for billions of years throughout all species of organic life, generation after generation, in this current form, otherwise you were not here now to confirm it.

Why do intelligences, organisms, and entire species develop in this exact form? Because this is the only way, since all life of the entire ecosystem changes simultaneously. There is only one way to develop, because there is not much developmental room left, since it has to happen simultaneously for all life, in this very tight space, which is this world. All intelligences are not interconnected only intrinsically, only within the organism, or only within the family, society, and species, because if left to interconnect freely, all intelligences interconnect with all intelligences of the environment, and in this manner, the entire comprehensive development taking place is interconnected itself, and it is unique.

Furthermore, since all development takes place through inner needs and feelings coming from Life herself, all comprehensive development is already part of Life herself. If the human society and the entire Consensual Matrix constrain you to remain underdeveloped and at the first servitude level matching coincidentally the level of the entire Consensual Matrix, this interferes with Life, you stop fulfilling your natural needs and meanings for Life but you fulfill only your artificial, consensual needs and meanings for all tyrants, you fail fulfilling

Life, you cease to exist for Life, you are dead for Life, and you exist only on behalf of all tyrants of the Consensual Matrix. This happens with everybody throughout the dark ages, while everybody is proud of this entire consensual achievement.

Intelligences are not individual living beings, but they are systems of intelligences themselves, resembling societies and civilizations, but not exactly individual living beings. While individual living beings are composed of zillions of living beings, realities, and forms of life themselves, resembling intelligences, communities, and civilizations. At the moment of conception, there are samples of your body and cognitive system sent from one generation to another, with parts of all your primal intelligences in it. Since your organism and cognitive system resemble to communities holding individual intelligences and individual systems of intelligences, samples of these resemble to colonies sent far away to start new civilizations. Yet since these colonies contain all the specialized intelligences that know exactly what to do and how to behave under all circumstances, similar to the rest of intelligences left in the parent organisms, now the organism of the new generation behaves similarly, while containing the same communities of intelligences that are still living within the parent organisms.

This is the case because life is always lived by the ionic form of life found at the base of the entire electromagnetic type of life, while this is the one reproducing, not the forms of life above. This is why all life reproduces starting with the ionic form of life, because all life here on Earth is lived by the ionic form of life and for the ionic form of life. It is similar while eating, because the entire food is broken down through digestion all the way down to the ionic form of life, assembled later on throughout cells in all cellular components of the molecular form of life, then in all cells of the cellular form of life, and then through reproduction, in all new organisms of the organic form of life.

Life is always lived in communities, made by ions, molecules, cells, and organisms, while all these communities have at their base the ionic form of life. Life reproduces through division, either through cellular division or through the division of the cognitive system into specific colonies of intelligences meant to form new organisms, and this is how organisms reproduce. Even when life is possible in different forms of life, made of different forms of life, it always divides in the same manner, on behalf of the ionic form of life, regardless if it is made of ions, proteins, cellular components, cells, organisms, nations, ecosystems, or realities. They always divide into colonies to send them elsewhere to restart and reconstitute the entire cellular component, cell, organism, society, nation, or reality, because this is the only thing that the ions know what to do, while all intelligences have their roots in the ionic form of life, this is what they know, and this is what they do.

As we keep noticing throughout all models of this book series "Human," life seems to be lived on behalf of its lower forms of life more than on behalf of its higher forms of life. All developmental needs come from all forms of life, lower and higher, meant to reassemble and redevelop life as it was before, and take it from there to develop it further, because all your needs and meanings come from your intelligences, while all your intelligences are systems of intelligences having their base in the electromagnetic field, in the lowest form of life standing right on top of the electromagnetic field, which is the ionic form of life.

More precisely, all your needs and meanings are meant to tend to the entire organism and to the entire human family and human society, while they always come from within your cells, because all your intelligences are systems of intelligences formed by all subcellular intelligences of all your cells, which are the actual base intelligences of the ionic form of life. The ionic intelligences tend to all molecules, cellular components, cells, organs, bodily systems, families, communities, human society, the entire ecosystem of Earth, this entire reality, and this entire cluster of created realities, all the way up to Life herself. This is why we could not find the hormone of life, the cell of life, or the bone of life as an essence of life, because life was in everyone and everything similarly, coming from within the ionic form of life sitting at the base of all life here on Earth. Yet since the ionic form of life resembles more to a community or to a society of living beings than to an individual living being, it seems that life itself comes from all individual living beings consisting the ionic form of life, which are the actual ions oscillating lively and intelligently in the electromagnetic field, while this must be in our definition of life.

Therefore, as you study the development of life at the level of the organism and of entire species of organisms, you must always consider it coming from within, from the base of life, all the way down from within the ionic form of life, made possible by all inner intelligences having their roots directly in the electromagnetic field. This is why all needs, feelings, knowledge, and meanings are correspondently similar from one form of life to another and then from one class of life to another all the way up to Life herself, because this is what the base intelligences of the ionic form of life know what to do, while they do everything in their own needs and meanings, one form of life after another starting with their ionic form of life. This is why brains have the texture of skin, the skin has the texture of cellular membranes, while the cellular membrane has a plasmatic appearance, because these are the base conscious ionic intelligences of the electromagnetic type of life while making the entire life on Earth possible, since all life on Earth is of the electromagnetic type of life.

Similarly, the cornea was a scale or a claw, and so was the hair after it was feathers. All adaptations that we see in the organic form of life look more as homemade appliances than rigorous improvements, meaning that the developmental intelligences of the organic form of life are not based at the macroscopic organic level of this world, or at the macroscopic species level, but they are deeply within organisms and cells, at nanoscopic levels, while they do everything in the macro world exactly as they know best in their micro world and nano world, within the ionic form of life. It is similar with you, because you also behave in your family as you know best, which might be different than what your family needs best.

You might not understand this statement if you always assume that what you do best in the family is exactly what is best for your family, yet this is the case not only in the family, but in all cellular components, cells, bodily systems, organisms, communities, societies, and civilizations, because you could have always come up with more than a horn in your eye as a cornea or more than hair on your body to protect your skin, yet this was what your developmental intelligences could do best, and therefore this is what you now have.

As stated previously, it is only through large numbers of intelligences and through their successful interconnectivity that cognitive systems achieve second level intuitive intelligence. You might expect intuitive intelligence to start with cats and horses, but we find it even in prokaryotes and in their cellular components, mostly because they are filled with zillions of intelligences themselves, and because prokaryotes live life in entire cultures of billions or trillion of prokaryotes. You will not find isolated bacteria for too long, because these divide in mass and you end up with billions of cells and more, as long as resources last. Even if prokaryotes as bacteria do not form organisms as eukaryotes do, each bacterium from the culture is specialized, living life for the others and for itself. While the others do the same, as though they were an entire organism.

Life never lives individually, but within entire classes of life and within entire forms of life, regardless if it has a skin or membrane around it, or nothing at all as it is the case with molecules, species, cultures, tribes, societies, and civilizations. If you search very closely, you can still find communities of proteins with or without an RNA, living life normally in the environment without a cellular membrane around them, while interconnecting with entire cultures of bacteria and algae in this manner, continuously throughout the ages. Yet even these living communities of molecules have at their base the ionic form of life, or they are of the ionic form of life altogether, not of the molecular form of life. As you study these living communities of ions and molecules closely, you notice how they interconnect not only within themselves, but also with the other living beings and living communities around regardless of their form of life, with everyone possible, in the best harmonious manner, because this is how it is easier to fulfill their needs. Living harmonies can be so efficient under specific circumstances, that they reach very specific levels of resonance, from the second intuitive level to the tenth supreme level, called the music of the spheres. Even bacteria will interconnect with entire small organisms, while fungus will interconnect with plants, animals, and entire species of these.

Which living being exactly is more capable to survive? The more developed species survive better, since they are more capable. The most primitive species survive better, since these are not too affected during cataclysms, being too primitive, and always survive.

Survival is not about individual organisms, but about entire species. In order to study survival, subsistence, and development of species, you must speed up time, to see in this manner entire forests moving around the continent while chasing their specific niche, since niches always move around the globe, with all species following them closely. Because there are no invasive species in the world, since it is already their world. When you speed up time, you can see how specific successful species chase original niches or they adapt to new niche elements while discarding old ones, or else they go extinct.

This is why you need specialized intelligences, to tend to each niche element. While as you speed up time, you can see how birch trees are just as successful as deer, which are less successful than elephants, which are more successful than entire species of snakes, which are less successful than specific species of fungus. Since it does not really matter your place in taxonomy, but it matters how capable your intelligences are to take over newly available niche elements and therefore specializations, in order to perform whatever other intelligences are incapable to do. This is the case for all intelligences of all living beings of all species and of all forms and types of life, in all realities of the wider world.

Because it is not exactly the body to count, but the intelligences within. Life involves a great effort in order to improve its cognition, to make it intuitive and rational, in order to be able to predict future events and obtain new ideas on how to overcome them, achievements that we might underestimate throughout our study. Yet we do so because we continue to remain incapable to tap artificially into the raw elemental intelligences of the field, to allow our robots to become alive.

We are fast-forwarding the model, waiting for the falling rocks to occur in order to see if the very strong materials used in our robots withstand the impact, when a big flood happens, drowning the robots. Our robots can still survive underwater, who knows, if they learn how to swim. We fast-forward the model for centuries until the waters recede, to find the robots inert, totaled, decomposed into rust, and buried deeply in the dried mud.

What a dreadful event, destroying our entire community of robots, after doing so well. Yet we are used to the environment by now, yet this happens with all species, since none seems to survive more than our robots do. Are our robots already alive, matching well the living beings of their environment? Yes and no, depending on what you consider our robots to be. The metal of our robots is already corroded, gone, yet it is not exactly lost or inexistent, since life made well use of all rust, plastic, rubber, and of all molecules, including them within amino acids, proteins, and enzymes of actual living beings. Everything is part of life, of natural life. This is how the actual living intelligences of the field end up tapping into our robots to use them from now on as part of Life, and not the other way around with our robots tapping into the living intelligences of the field to become alive. Life herself took our little robots as she always does, making them alive molecule by molecule, as physical parts of other living beings.

Life

Our robots are alive, at last, recycled by life itself. They will actually be alive endlessly now, caught in the entire intelligent development and interconnectivity taking place within a multitude of species and living beings, while this is natural life, of the second intuitive level and higher.

We are successful at last, job well done, but not exactly, since we consider our created living beings to be more than the sum of their parts, but to include their intelligence, behavior, consciousness, achievements, and development. This is how all extinct species are still alive in everything currently alive, through their own achievements and therefore through their own successful specialized intelligences that are always in us.

We continue our iterations, yet this time we decide that their recording devices were too small, and this is why our robots had failed to predict the big flood, through ignorance alone, since they had shorter floods before, and they had survived well. Therefore, we increase the capacity of their mechanical brains, and we restart the model.

Our new robots are smarter, it takes them less time to do their specialized chores, they use less materials, and therefore they have more time now, time that they still use saving energy and not learning. They manage to cope fully with the environment, they learn and discover new technology as needed, only to perish, sadly, in a different cataclysm, throughout the following century. Life took them again.

We make our robots more resilient this time, increasing their level of technology. Additionally, we install backup systems, everything with the same dreadful result, total mechanical failure, and inevitable destruction.

Why does this happen? Is our model failing us? Yes and no. Most of our settings are still set on normal, our model is always accurate, and now these are the results that it states: most communities, societies, and civilizations go on for about a few millennia at a time, while confronted with a normal set of random environmental events here on Earth. Yet all living beings and all species die eventually, throughout more dreadful global events. Our results are even backed up by history, biology, geology, and taxonomy. Yet humans always survive all cataclysms, even though they lose their societies, nations, cultures, and civilizations in the process, to reborn afterwards as the Phoenix bird from their own ashes, to start again.

We notice that humanity has the same faith as our robots, not exactly because there is no other way throughout the harsh environment of Earth, since the other species survive well, just study flies to see it for yourself, but because they live life at the first developmental level, one in a mechanical form, and the other one in an artificial consensual manner during dark ages, with none genuinely alive and none intelligent, causing our robots and the undeveloped consensual humanity to fail. This is the case with our robots because we cannot make them alive as we always intend, while for the currently undeveloped consensual humanity, there are a multitude of entities and civilizations up there that want humanity to remain underdeveloped and out of their way, not to have to share their advanced niche with them even endlessly.

Our robots cannot survive for more than an entire Earth age or two, and we might have to discard them, at last. Yet if humanity cannot survive from one age to another, then other species will, and once they become intelligent, they might succeed in maintaining Earth in intelligent golden ages, at last.

Why do intelligent species fail, if they are intelligent? The Consensual Matrix targets mostly intelligent species and higher, because they are more profitable. However, if you are a third level intelligent species or higher, and if you cannot protect yourself from the first level Consensual Matrix, it means that you are not a third level intelligent species or higher. While in general, once you start taking drugs and once you rule tyrannically, you decay to the zero and first levels, below the level of all species of Earth. Yet since you monopolize the third level intelligent niches of Earth from your addicted and tyrannical level, you ruin the world irremediably. Just study humanity closely to see it yourself.

Even during milder global cataclysms, humans go extinct, with the Masses to go extinct first, and then with the

Brotherhood and the Elite to follow after lasting a few generations more throughout tight underground habitats. Even cats and horses survive better than humans do, while rats and roaches thrive during cataclysms and after, expanding their reach to span the globe, which is successful survival. Bacteria will outlast flies and horses, which are the most resilient among all species. Because as you notice, there is no actual death in Life and in the wider world, but life only shifts from one form to another downwards when it fail, and upwards when it is successful. Therefore, since bacteria is of the cellular form of life, it lasts longer, even after all the organic form of life is gone, with humans, cats, horses, flies, and roaches included.

It is a stereotype to consider species striving for development and evolution, with humans to be the most capable of them since humans have houses and shoes. Humans are the most capable and the most developed, but only at their intelligent level sober, not intoxicated at the zero level. You always fill up a specific niche through your characteristics, while only one of these characteristics is your development, while only when you are intelligent you are successful as a human being, never otherwise. Therefore, whenever your ecosystem needs you at your intelligent level, but you are not there because you serve tyrants and you take drugs as everybody does throughout dark ages, then this is how you go extinct alongside most of the organic life of Earth. While drugs never make you feel good but dreadful, as you know it firsthand. It is the same with tyrants, because they never do what they promise, but they always lie, while they always take advantage of you. Therefore, all dark ages are not good for humanity and therefore for Earth, but they are even harmful, regardless of everything that you might assume.

We notice a specific downgrading of life to its lower, more resilient species with each significant change in the environment, or even down to lower forms of life if the change if the environment is drastic. Again, we observe lower forms of life being privileged against higher ones throughout dreadful environments. Whatever the case is, life is never lost during major cataclysms and major changes in the environment as science claims, but life is only lived on lower species, lower levels, and even on lower forms of life. The organic form of life has the ionic form of life, molecular form of life, and cellular form of life at its base, with ions, amino acids, proteins, and ARN chains of proteins at its base, and this is not organic life, but ionic life and molecular life. Whatever the case is, whatever the change in the environment is, life will always occupy all niches of the new environment accordingly, even in a different, lower form of life, or even in a different type of life altogether, however it is more adequate for life.

There is no death for life, regardless if humanity remains undeveloped, because there is only death for humanity. It is similar with all unsuccessful intelligences to cope with all changes in the environment, because life manages eventually to cope under all circumstances since there is never a death for life, but it is only the death of those specific unsuccessful intelligences, becoming inexistent from then on indefinitely. Study closely the very old vestiges of Earth, to see how they are left behind by very old unsuccessful civilizations not human, but of other species, even reptilian, now gone indefinitely, through similar lack of development and lack of success.

We want our robots to live for more than a few millennia at a time, since we want them to survive these major destructive events, which are relatively mild compared to what is coming eventually, even more destructive events. What do we do? If learning and development always count, then we should increase their learning parameters. We are fortunate because our robots cannot take drugs, because they do not have feelings, since feelings relate directly to the second level intuitive cognition, absent from our robots. Because if our robots took drugs, they failed even more often than the current undeveloped humanity. Yet since they cannot take drugs, they actually last longer than the current undeveloped humanity, helping us better in our study of life.

We restart the model with a learning capacity set at two

hundred percent now, which means that our robots are twice more willing to adapt older solutions to new problems. The model starts, and what our robots do, they tend to solve everything around in every way, many times unnecessarily, wasting their time and resources. Furthermore, they tend to use more their new solutions to old problems, and not the old, provided ones, causing them to fail by default, compromising their entire survival. Our learning robots last exactly twenty years before they break down irremediably, which is still good.

Why do living beings survive naturally and our robots do not? Why do humans survive better than our robots? Humans are systematically modified throughout dark ages to be obedient, hardworking, and extremely resilient, and it is the same with horses and with all domestic animals. Humans do not adapt physically anymore to the environment, but they adapt the environment to them. Which means that we cannot exactly use humans as a reference for our model with robots, yet we can use horses. The question here is how to make our robots survive better than average communities of living beings, since as seen, learning, strength, and excellent technology do not seem to make a difference.

What we did wrong, we isolated our robots in a private setting. Life never isolates her species, but life makes them interconnect with each other in the most harmonious or dreadful manner, depending on circumstances. It certainly hurts life when the wolf hunts the sheep, since valuable, healthy organisms are lost only to feed others. Yet this keeps both the wolf and the sheep around, diversifying life considerably. This is what our model lacks, interconnectivity of all kind, as diverse as possible, not only learning and development meant to cope with the environment. Life is intelligence, interconnectivity, and the wider world as one, not only intelligences as we highlighted so far.

We are peaceful researchers, we tend to live life in peace at the third intelligent developmental level, and if we are not careful, we lose the necessary condition of life forcing us to survive all changes of the environment. As intelligent human beings, we engage only in win-win circumstances, while making sure that everybody survives, subsists, develops, and prospers throughout all environmental conditions, good or bad. Communities of people are more likely to survive during crises, since people help each other throughout all intelligent human communities, part of the comprehensive third level intelligent human interconnectivity. However, there are specific changes in the environment sometimes creating niches that allow only very small communities to survive, as only families or only individuals, for shorter periods, before they start interconnecting intelligently once again. This is what challenges intelligent humans, both the lack of intelligence and the lack of interconnectivity.

We change our scenario accordingly, and we place our robots in a beautiful tropical island setting, where we let them survive and compete with all the wild animals present there, including snakes, ants, wasps, large herbivores, and even small monkeys.

From the start, all life from our new ecosystem defends its territory by challenging our robots, yet our robots do their usual chores, never impeded by the wildlife. They do their little tricks and invent the necessary technology throughout the years, while all animals learn to tolerate them. The following generations of animals consider our robots as casual part of the environment, part of the neutral environmental conditions and part of a separate food chain, so there is never a problem.

Our robots learn more on the island, since this environment is richer in events. However, our robots learn everything up to the level required by that specific environment, and when the volcano erupts one thousand years later, they do not have a boat ready to take them away from the island, they are not able to fly away, nor are they able to move their habitat underwater, since they had never predicted anything so unfortunate. Therefore, our robots break down and cease to exist, alongside all the wild animals of the island, which is dreadful.

What our robots lack is higher curiosity, the need for

intelligent knowledge, which is a relatively higher need, a third level intelligent need, because we had embedded only first and second level needs into our model, but not third level intelligent developmental needs. Why did we do so? Because we wanted to build a living being, anything intuitive, not directly an intelligent living being. Because we are told by the current science that the nonliving evolves into the living first, then living beings evolve into intelligent human beings, and now we are still waiting for our robots to evolve naturally into living beings, and, if possible, into intelligent living beings. While they always tend to remain at a very low mechanical first developmental level, which is not sufficient for the challenges that the environment poses over a very long period, and it ruins our robots. Yet this is what also destroys the human civilization, in about the same manner and in about the same time.

Why having this issue? Because we have already discarded the theory of evolution for being insufficient, and now it seems that we have to discard the entire current science, for being similarly insufficient. Everything happens because it seems that the living does not exactly evolve from the nonliving, and furthermore, it seems that not all species evolve into reasoning intelligent living beings. Probably because they do not have to, because their niches do not require third level intelligence. We already know that development relates to interconnectivity at the level of all natural intelligences and at the level of all living beings, yet how can we apply this in our model of a created living being, which does not even have natural intelligences, but only some mechanical punch cards or a rudimentary CD encoder?

We decide now to place our robots not in deserts and tropical islands, but right in the middle of Europe, currently, and let them interact with the toughest of the toughest living beings ever, humans themselves, as raw as they currently are in Europe.

Once in Europe, our robots are observed closely from the beginning, as they are never left alone, even by the media.

Equipped with brand-new computers, our robots start their European existence chopping wood for a specific man from Belfast, who they venerate now, for no specific reason. For him and for his family, our robots replicate themselves, they work all day and all night, they clean up streets, they work the fields, they load and unload boats, they do willingly any work there is to be done, all for this specific man from Belfast, while he and his family become very wealthy very fast.

This interferes with society, since wealth is distributed to all families in a prearranged well-agreed order called legacy. Measures are taken from far above, and very shortly, this lucky family is compromised, ruined, and banished altogether. Our robots receive some rights, and are suddenly free to decide who they can work for and who they can venerate. Our robots think very well, and decide unanimously to continue to venerate that man from Belfast and his entire family.

Centuries pass, and society accepts our robots with open arms, as slaves, since every family can buy one of our robots now, being very cheap, as they even self replicate. This is how our robots learn less and less, only what humans allow them to learn, mostly related to how they can do a better, faster job throughout all farms, factories, and households. Our robots become the disposables of the human society, similar to the Masses of the current society, because the current Masses are very cheap to acquire, as they even reproduce themselves, while they are similarly eager to serve, for the same twenty dollars in benefits. Furthermore, if you only point to any man from Belfast, New York, Moscow, or Beijing, they will venerate him and his entire family even endlessly, or until you point to someone else, to venerate that one from then on. Until you point to someone else, then to another one, then another one, and another one, one dark age after another, because the current Masses will always serve, for the same twenty dollars. Just give them the twenty dollars, and they will always serve, while this should also be in the definition of the human life.

Consequently, our robots improve less, and they never

receive an equal human status, not even a living status, but only a first level corporation status, since they are inferior to men through intelligence, awareness, and abilities. In the end, our robots die in servitude, along with the entire human race, when a larger asteroid strikes Earth casually, leaving only the upper social class behind, crawling throughout caves and tunnels for two more years, while perishing gradually, incapable to fulfill their own needs without servants, as cooking for themselves, dressing themselves, and making their own beds.

However, as you study the second level animal ecosystems of Earth, you find them in perfect harmony, not too much tyrannical and in competition. Even throughout food chains, species do not struggle dreadfully for survival by eating all the chickens in the first day in order to be able to survive in a competitive world, but you always tend to the wellbeing of the entire chickenkind if you happen to be right above them in the food chain, since this specific harmonious behavior saves you both from extinction, your own kind and the chickenkind similarly.

We place our robots in an abandoned city this time, directly in Chernobyl. We place not one, but four different casts of robots: the strong, the advanced, the smart, and the fast-learners, in four different parts of the city. We increase their coefficient of competition, we start the model, and we watch them closely, very eager to see what happens.

Our robots cooperate, forming a larger community, despite of their difference in abilities. Yet they do so only for the remaining of the century, until they finish melting everything for parts, and ran out of metal. That is when the competition starts on all levels and in every possible manner, going all the way to capturing and melting down random robots, only to use them for their raw materials.

What is different now from the previous runs of the model is that separate communities of robots fight for the supremacy of resources, and they refuse to share resources and information with the rest of the robots as we had instructed them to do. Knowledge is treasured and sought the most now, just as we wanted it from the beginning. Yet the fight for more knowledge takes place in a very brutal manner, unlike what we had actually intended, since we wanted all learning and coping with the environment to be done entirely in a peaceful manner, and it never happened.

Does this mean that our robots are bad? Does it mean that we are bad as creators? Is our model evil now if we introduce sources of evil in our model as competitive means of living? Is our model the source of all evils? No, since nothing in Life and in the world is evil. You can certainly strive to be good in a lower level environment, but you fail, and you die away, because your specific developmental mode of life is inadequate in an undeveloped environment, being too high. Because you always have to match your mode of life with your environment, otherwise you fail, and you die away. Yet at their third intelligent developmental level, humans should always be able to change their environment on their behalf, developing it entirely to the third intelligent level, while forming the third level intelligent human society, since you always have third level developmental environmental needs and meanings in you, and you have to fulfill them in every manner.

You live your life within a multitude of environments simultaneously, as the natural environment, cognitive environment, social environment, and consensual environment. Harsher natural environments might require higher modes of life to succeed, as harmonious, cooperative, and humanitarian modes of life. However, harsher natural environments, as natural cataclysms, can also shift societies into lower, offensive modes of society, while you cannot be humanitarian throughout offensive modes of society, or you die away, because those around always kill you either directly or implicitly.

Capitalism itself is an example, since it allows the rich to reproduce more than the poor, and therefore the genetic lines from the bottom social layers have no chance to survive. They are worse than our robots. This happens with all genetic lines, one after another, social layer after social layer, from the bottom of society up, which is called genetic eradication. Which means that our robots survive significantly better than the humankind, because the current humanity remains underdeveloped, at the first ideological consensual level, below our robots. This used to be the case until now, because once we enhance interaction in our model, everything can happen, since interaction itself is either harmonious or competitive, and we never know what our robots choose. If they choose harmony, they might last longer. Yet if they choose social competition instead, they help us learn more about the current consensual hierarchic competitive human society.

We run the model again, and in only decades, our robots invent not only the CD, but they invent the computer just as well, entirely. Again, nothing is invented directly, since our robots lack full intelligence and intuition, yet they duplicate the technology used in their power invertors, creating a different version of a power invertor, capable to store and process data, just as a genuine computer. Our robots are not even intuitive enough to invent these new, better computers and to equip themselves with them, but they adapt an entirely new specialized robot equipped with the modified, oversized version of a power inverter, and they use the new type of robot entirely as a computer. While it is still a robot, an entire robot, as fat as it can be, now used as a computer. These specific new types of robots are specialized now in the storage of data of all kind.

One century later, this important technology is part of every robot, yet each community still has its own giant specialized robot, used to store permanent reserves of data for future replications, with the rest of the robots keeping in their computers only what they need for their usual tasks.

Robots form communities of every kind now, mixed communities, and even communities including several communities of robots along with individual robots. These giant robots are used to store permanent information. It will not be long before the drastic miniaturization, when you have communities of communities and individual robots within other communities. This is how Life folds upon herself to form newer forms of life as new organisms and entire new societies.

Why does it happen? Because life is not fully intuitive even at the second developmental level and higher as you might assume, since intuition is achieved through high number and through high effort, always through intensive diversification and specialization. In our case, the more robots we have in a community, the more diversified their intelligence is, and the more new ideas they have. Because through intuition, Life always obtains something out of nothing, as creativity, new ideas, new possibilities, choice, expansion, development, and improvement. Intelligent cognition is even better, since only intelligent cognition is capable to apply intuition to conceptual intelligent circumstances, creating an entire new mode or level of life, the intelligent mode of life, at the third cognitive level. In contrast, the second intuitive cognitive level is more empiric in nature, requiring larger numbers of intelligences and larger numbers of trials, while the intelligent cognition of the third level can be achieved even through individuals, and it can be very accurate.

Even rational intelligent cognition is based on intuitive cognition throughout the entire reasoning process, with the entire intuitive cognition being done within the inner mind realities of the individual, by the multitude of its own inner intelligences. Because as you notice, higher cognitive developmental levels are not only simple developmental abilities or characteristics, but they are entire new classes of cognition, developing and being instated throughout the inner cognitive worlds of the mind, and these classes of cognition are correspondent to the forms of life found in the physical objective world.

There is a difference between a more intuitive intelligence and a less intuitive one. The higher is the degree of intuition, the more capable a community is to find solutions and ideas, the more adapted it becomes, and therefore the more it survives by taking its resources from the lesser intuitive, lesser fortunate communities and individuals.

Everything in this model is similar to cellular life and cellular intelligence. The cellular community resembles a community of robots and machines, more than any other community of individual beings and classes, and it does so through its structure, behavior, individual and communal replication, thinking, development, and accomplishments. The cell is a genuine community of zillions of little robots, as ions, molecules, amino acids, proteins, and enzymes, along with larger individuals, and along with even other communities of smaller and larger individuals.

This was the case from the objective perspective of the physical body, since from the cognitive perspective of all intelligences carried by our robots, robotic life is but a simplistic version of natural life, lacking the intuition and creativity that living beings can manifest. While in the case of the cognitive, subjective cellular life, there are a zillion cellular intelligences thriving within all cells, specialized in everything that all cells need. It is the same with organisms, species, societies, and realities.

Why the resemblance between an imaginary model of life and real life? Because nothing in life is random, but only intelligently decided, in every way. The laws of survival determine everything, because the more experienced you are and the more knowledge you have, the better is your chance to survive not only the environment, since the environment poses irrelevant threats, but the better is their chance is to survive other communities of robots, and it is the same with humans.

This is how, for any form of life, once placed in a specific environment, you do not adapt and develop in any specific manner, according to your experience and stored information, but you develop in the most efficient type of life and form of life capable to cope with that specific environment, type of life or form of life that is always unique for that particular environment, while this is life. You do not fight and cope with your environment, but you become the environment, you integrate harmoniously in the environment, while this harmonious integration is unique, since it is the environment itself. Our little robots were not the environment from the beginning, since we had placed them there ourselves, this is why they cannot survive, and this is why they still fight the environment, including themselves. How exactly will they become the environment? They cannot.

Comprehensive development is never a probabilistic matter. Development, cognition, behavior, modes of life, and all achievements are always a matter of strong correlation and compatibility with the environment from the beginning, before species get to survive or go extinct. Afterwards, if species happen to go extinct, they do so because this specific compatibility with their environment is no longer the case, for various reasons, related either with the species itself or with the environment, but mostly because their niche is not available anymore. It is the same with the environment itself, since all environments are alive, and it happens to them the same as it happens with the species that they sustain.

I refer to this as the dance of life, when species do not exactly cope or compete continuously within the environment, but they always dance with the environment in a perfect synchronicity, in a perfect marriage, in a perfect harmony. While when you have harmony, you do not exactly have competition. This environmental dance or marriage can go through good times and bad times, it can be harmonious or competitive as it happens with everybody, since all individuals are in themselves entire forms of life and cognitive systems, they are entire environments themselves, and they always make it alive through their own means, through their own algorithmic, intuitive, or intelligent cognition, but not at all randomly, through radical competitive scenarios as stated by the current consensual science.

Life is always intelligent, Life already spans the wider world, Life is the wider world, which is her natural environment, and therefore Life does not fight the wider world in a radical survival scenario as stated by the current science, but Life dances harmoniously throughout life and throughout the world, in a minutely detailed arrangement of all her living beings, species, forms of life, cognitive systems, and entire realities.

Natural environments can change drastically every few billion years or so, due to astronomical circumstances, which refer to even larger environments that hold human environments, since this is always the case. Yet if an entire biosphere cannot adapt to the new conditions since the change is too drastic, then the entire biosphere does not exactly perish, but it morphs to a different class of life or to a different form of life altogether, in order to meet all details and all characteristics of the newly changed environment.

This happened on Earth over one billion years ago, when the sky cleared and sunlight touched the surface of Earth, allowing photosynthesis. The most efficient form of life under the newly changed environment was photosynthesis, with all the old intelligences still using chemosynthesis, through all possible chemical reactions feeding them, while the newly adapted specialized intelligences used sunlight and photosynthesis. This is what many living beings did and still do, with sunlight itself as a new element in their own niche.

How exactly did the newly specialized intelligences learn to use sunlight to generate energy? There are two main answers given by science. The first one is trivial: the living beings and entire species incapable to use photosynthesis kept using chemosynthesis or died away extinct. The second answer is erroneous: there are always genetic errors occurring randomly during transcription and translation, always capable to generate the multitude of adaptations that life undergoes throughout its entire evolution. Which means that the entire development throughout life is made randomly through errors, with all the paid scientists of the world counting in millions never embarrassed to state so, while all teachers and professors teaching this are never embarrassed either.

How exactly are the newly specialized intelligences capable to conceive and develop the entire chloroplast within many prokaryotic cells? It is more difficult than you might assume.

The old prokaryotes did not simply warmed up in the sun directly, and now all cellular intelligences learned to warm up and simply used the newly available energy, because the entire prokaryotic cell is only several times larger than the wavelength of the color green, while the color green is the most powerful of the entire spectrum of visible light. All cellular components are significantly smaller than one wavelength of light, making light and all its energy pass straight through them, and cannot use it.

Again, how were the newly specialized intelligences of the chloroplast capable to conceive the entire chloroplast from nothing at all, since sunlight had never touched the surface of Earth before? They used the optical phenomenon of interference, and this was how they managed to warm up in the sun. More precisely, they placed proteins in larger arrays at very specific distances one from another in a well-determined pattern, and in this manner, they managed to block sunlight so it does not go through them anymore, using it by converting it to charge and therefore energy, and then loading it into ATP molecules. They already used exothermic chemical reactions throughout chemosynthesis to produce charge and therefore energy and ATP molecules, and now they simply used the new method to block sunlight and used it as energy, further converting it into charge and ATP molecules in the old manner used through chemosynthesis. While furthermore, they modified the chloroplast into melamine within skin to block sunlight in order not to harm the skin or bark, while later on, they modified it in a retina to block sunlight again in order to perceive light and images, while forming the eye. They also used a claw as a lens or cornea in the eye, forming the eye. Homemade appliances.

We have the same intelligences in the skin and in the eyes used previously to make the chloroplast, this specifically arranged array of proteins using the physics phenomenon of optical interference. Yet they also had to hold themselves in place very precisely throughout the entire optical interference, so they used a harder transparent substance to hold them steady in place.

How exactly did the protein intelligences know to form this entire organized array capable to block sunlight through the optical phenomenon of interference? The entire prokaryote was still capable to block sunlight at least slightly, since prokaryotes measure in micrometers, while the wavelength of the color green is about one half of a micrometer. While within the prokaryotic cell, a multitude of ions and proteins were capable to block sunlight together and therefore they were capable to use it as warmth and energy but only if they were placed at a specific distance one from another, part of a very specific pattern within an entire array. Furthermore, they placed as many specialized proteins in this specific optical interference pattern, holding them there still in a harder transparent substance, so they stay still and do not drift away. Yet mostly in order to be able to transfer the newly harvested sunlight energy to be transferred and converted into charge and then into ATPs. While as you notice, everything was made possible by an entire cellular interconnectivity formed by zillions of inner cellular intelligences. This is how we have chloroplast intelligences ever since, in all green bacteria, algae, and plants, which is basic living cellular intuition.

Do our robots become intuitive, even only as a community, and even only in the most proper environment allowing intuition? The answer is no, sadly. This is not our fault, because you cannot match Life by creating life yourself. You have to be alive first, in order to become alive. However, we can continue our model as though our robots develop their intuition, only to be able to continue studying intuition itself, along with the entire life, at the cognitive, physical, and interconnective level.

Do not regard our robots as simple toys, because if I called them drones instead of robots, with all machines and robotic arms from all factories of the world working automatically while assembling all drones, and with more drones gathering resources and building all the necessary drone parts, our model of a created living being becomes very credible. While all drones are normal robots, regardless if they are called drones, machines, little robots, or robotic arms.

Imagine entire nations at war using our robots, machines, robotic arms, and drones to make more military drones, automatically, to fight in the war automatically, by our current technology. You can have entire armies of drones fighting entire armies of drones for some time, after humanity is long gone, because military drones are capable now to replicate themselves freely, even indefinitely. Yet again, the current computer technology cannot ascend from the first algorithmic level to the second intuitive living level, failing exactly here where our little robots had failed to become alive. Because you cannot tap into the raw intelligences of Life to become alive, since these intelligences have to be in you from the beginning. Yet we still continue the model our little robots shortly, only to be able to mimic life, while still learning more about life, because we still want to understand life standing right on top of the electromagnetic field, while forming the entire electromagnetic type of life, with the organic form of life included.

Wireless technology happens now in our model with little robots. Wireless signal is electromagnetic in nature, closely related to electric and magnetic field, forming the field in general. Our robots 'invent' networking itself, yet they amplify as much as they can a specific interference that they experience when they are too close to each other coming from their power invertors, and so they transfer data wirelessly within all their communities. Our robots have already built a strong metal shield around their community to keep them secure from invaders, and now they have to build a similar shield to keep their wireless signal confined to their community and not reveal information, and they do so by simply grounding permanently their metal shield.

This is how the first solar flare finds them, tucked well within their own communities, well shielded from the outside spies. Therefore, this is how they survive their first medium level cataclysm, happening once every few thousand years, while all robots caught outside their communities were fried sadly, along with the entire electric network of Earth. All electronics, appliances, computers, smartphones and databases that humans had by the zillions were fried, plunging humanity back into darkness for decades to come. While from the way our society behaves currently, humans certainly remained in darkness for much longer, under sheer tyranny and continuous servitude, while producing yet another unnecessary dark ages.

Yet our robots survive, they thrive, and they leave a small note for the future replications in their permanent memories, about how to detect and survive solar flares. Decades later, elaborate encryptions and powerful signals allow every member of every community to roam everywhere and still be in close contact, wirelessly.

You cannot develop this new wireless technology by modifying progressively a power invertor, since it is possible technically, but you need intuition in order to do so, and you cannot have it by using computer technology. Because all computers are only a simple abacus altogether, only using electricity to function. While about surviving solar flares inside Faraday cages, yes, it is possible. While about solar flares marking the end of the human civilization, yes, it can always happen, mostly since all tyrants prefer humanity in a medieval type of dark age, lacking electricity altogether. While once you mark electricity itself as evil or as the devil, humanity will always avoid it, even by killing each other for mentioning the word electricity.

The lack of materials had changed our robots entirely. Size matters greatly now, since the more numerous a community is, the more experience you have, the more you learn, and the better you survive. Members of a single community count in millions now, most of them as small as bees. These are the gatherers and the data processors, while they still have in their fortress community giant information storing robots, giant power supply robots, modem robots, replicating robots, and weapon robots. The shielded fortress holding an entire multi-leveled community moves around now and drills for metals and oil, while it moves around using legs but not wheels, since our robots are incapable to invent the wheel, yet. Even so, they still do not adapt directly dozens of legs meant to carry the giant fortress, since they do not know how, but they simply build a giant robot to hold all of them, giant robot that already has dozens of legs everywhere, becoming now the giant fortress robot.

If you saw this giant fortress walking around, as large as a high school, and acting more as a bee hive while drilling for oil and mining for metals, you could assume that it was one giant robot, one individual robot, since it behaved exactly as one being, while gathering resources and while acting and reacting to the environment on its own, even showing awareness towards you and those around. Yet this giant fortress is a community of communities including other communities of robots, all mixed there inside, in every possible manner. The great majority of robots forming communities never go outside, living their entire existence inside, many times performing only one task, until they are recycled. Their interior environment is controlled, at constant temperature, lacking humidity and oxygen, in order to avoid corrosion. Because now, corrosion, for robots the size of a regular mite is fatal.

Why is life folding upon itself in this extremely tedious manner? It does so only to keep the largest number of individuals in the most confined place, keeping them as simple as possible, while using the least amount of energy and resources, in the most organized manner, which is a great achievement. Yet why having zillions of robots, and not a very large unique one? Because a zillion different individuals have a zillion different minds, thinking in a zillion different ways simultaneously, and technically, being capable of generating a zillion different ideas, simultaneously, while this is genuine intuitive thinking, at least. This is very helpful, since our robots do not store only final solutions for a later use anymore, but they store every possible information about the outside environment in their common brain, a genuine replica of the outside world, very similar with the one that you have right now in your intelligent conscious mind spanning the cortex, mistakenly believing that it is the outside world.

The entire walking fortress is a multi-leveled organized being, which can be considered by all standards as an organism. Studying the etymology, to see how an organism is an organized living being, while the communities forming it are also organisms, but of a lower form of life. The individual robots forming communities are simply individual beings.

Starting with the innermost beings, the innermost robots are considered individual beings. The community that they form is a living community, which is a living being of the first class. These communities of the first class are together with several other similar communities, and along with larger individual computers used in data storage, they form larger communities, which are beings of the second class. All similar second class communities, tens of thousands of them, along with larger computers used for data storage, power supply, and network system, form the fortress, which is a unique, ultimate community of the third class, behaving exactly as an individual being, the giant walking fortress, a third class being, or an organism. This is you, since you are also a giant fortress, living among giants, a fifth form of life of the electromagnetic type of life, an organic living being, while everything stated above also applies to you.

Biology, along with the theory of evolution, state that you are alive, while your body is formed of trillions of nonliving cells, eukaryotes. Each cell had been formed that way randomly, and therefore you had evolved to look as you do, making the transition from the nonliving to the living, which is not true. Because there are zillions of individual living beings forming you in very distinct, very elaborate manners, since all cells and all cellular components are alive. Furthermore, all cells give you life as an entire organism, and not the other way around. Similarly, all living human beings give life to all their families, communities, and societies, making these alive, and not the other way around. While all cellular components give life to the entire cell, exactly as they take it continuously directly from the raw electromagnetic field themselves. Because continuously, all life is lived at the level of all cells and below, at the level of molecules and ions, deep down in the raw electromagnetic field making possible this entire electromagnetic type of life that we are.

You are a community of seven trillions living, intelligent, and developing eukaryotic cells, forming your physical body entirely, while giving life to your entire organism, through their own life. Additionally, you have seven trillions prokaryotes forming the flora of your intestines, helping you digest. However, each one of your eukaryotes forming your body is a community of cellular components in itself, of the molecular form of life, which means that it includes several other communities, other cells, twenty times smaller, which are also prokaryotes, which are the old cyanobacteria.

Life had folded upon herself to give birth to you the way you are, since nothing is random about you, but very precisely elaborated. The cells inside your eukaryotic cells are called mitochondria, they are certainly alive, they are prokaryotic cells similar to any microbe, responsible with converting food into electric charge and energy, similar to the chlorophyll and chloroplast within plants. We find individual beings within mitochondria, proteins mostly, all working together while converting glucose into ATP molecules used as instant source of energy and instant source of electric charge by all cellular components everywhere within your body.

Technology improves, and centuries later, most of our robots, zillions of them, measure in micrometers, and this is about the size of a cell. They float, fly, and hover around within their communities, very tightly packed together, with some of them able to go outside for various tasks. It is amazing to see technology repeatedly improving and rendering them so small and so numerous. What it is even more amazing, they stay connected, they communicate with each other and with the server, filling up the airways with a cacophony of electromagnetic waves on all frequencies so intensely, that they use the same signal as power supply, propulsion, flashlight, heating, and weapon.

Humans can perceive and understand this cacophony of electromagnetic radiation, electric, electrostatic and magnetic fields formed by our electronic devices everywhere around us as Wi-Fi, cell phone signals, TV and radio signals, microwave, blue-tooth signals, radar signals, along with all natural signals coming from the Sun, Earth, and from every cell of the electromagnetic type of life, which I simply call the field. We can understand everything as a continuous transmission among electronic devices, computers, servers, modems, and relays, going to and from each other continuously. However, our robots are the direct users of this highly complex electromagnetic field for one thousand years now, and by now, they understand the field not as humans do, but they understand it exactly as it is. Our robots never distinguish between transmission and reception signals anymore, but they focus on the objective, the direct intelligent modulated contribution that they bring to the field. Furthermore, our robots use tools now whenever they need, they use simple ions and small charged radicals systematically placed in the field, and by using the modulated field, they move them around in specific sequences, shaping and reshaping the field in this manner as they want. They can do the same in the field, moving in specific manners in order to shape and reshape the field as they want, while this is common life for all ions and proteins within all cells, in the entire ionic form of life standing at the base of the entire electromagnetic type of life.

Our microscopic robots can literally view this complex electromagnetic field directly, just as we see with our eyes all around us. Our robots sense more than visible light now, the way they used to do when they had a video camera. They can perceive most of the electromagnetic spectrum, and therefore they are able to distinguish specifically the modifications that they bring to the electromagnetic field with their own computers. For them, the electromagnetic waves are not seen as transmitted and received waves, but as processed and unprocessed field and radiation, since our robots act directly on the field as they wish, and alongside billions of other robots, they think, exchange ideas, repair ideas, assimilate new knowledge, and more importantly, they plan future projects, doing everything only within this field, within this extraordinary wireless common mind, in genuine thinking models of systems, objects, subjects and events from the real world. Because at this level, our robots do not have to undergo anything physically by trial and error anymore, as they used to do when they were larger, since now everything is perfectly planned down to the slightest detail of the field, even for new, elaborated solutions, all imagined, planned, and modeled in advance within their individual and common minds.

Our robots work now towards adapting a new technology, a new step in data processing. The problem is that our robots cannot miniaturize anymore, since their computers cannot become smaller. This holds them back from accommodating more individual computers within communities, and therefore from miniaturizing even more.

You hear more about nanotechnology, about nanobots, capable to enter your body while repairing, curing, and rejuvenating you, but this can never be done, or not as science promises. Yet there are many lies and promises coming from science, since this is not the first one. Let us study this case closely, since it relates to what our robots try to achieve right now.

It is easy to imagine miniaturized drones measuring in nanometers, as science promises, since it is easy to imagine them replicating each other so there is no need for us to build them, mostly when they are so small and so many. Now in our model, all our robots measure in micrometers and not in nanometers, because they are made of motors, power supplies, and computers. However, since matter is made of molecules measuring tens of nanometers, you can never miniaturize any of these parts under a specific number of molecules, since they lose their shape and characteristics. You cannot simply miniaturize your car to the size of nanometers, since molecules themselves measure in nanometers, and you end up with a blob of one hundred molecules but not with an actual car, since one blob of molecules cannot even give you the shape of a car, but only a blob.

You cannot have robots formed of dozens or hundreds of molecules put together, since that is a simple chemical compound, but not an automatic robot. It is impossible to have motors, wings, wheels, robotic arms, lights, and video cameras while measuring a few nanometers, since you need a zillion molecules to make all these possible, while you can put together only a dozen molecules at that very small scale, ending up with an ordinary chemical compound, never with a drone measuring in nanometers.

You cannot have the two or three molecules that you find at the scale of nanometers to put them together in order to form your highly sophisticated nano robot, because molecules themselves are already the size of nanometers, and now when you place them together by the zillions to make drones and robots, you end up with large systems of molecules measuring in centimeters at least, not in nanometers as promised by science, small enough to enter your cells by the billions in order to rejuvenate you. Therefore, nothing can make sense at the very small size of nanotechnology, being only two blobs of molecules put together, while these cannot actually repair cells to make you healthier and younger.

Atoms are only a hundred times smaller than a nanometer, comparable to the size of a nanobot promised by science, which makes nanobots made of such a small number of atoms, that you can even count them. While these can never do all tasks that science envisions, or at least not at that very small scale. Nanobots have to be very small to enter the cell, while they still have to look and behave as normal drones and normal robots, which is impossible

Furthermore, at the scale of nanometers, nothing makes sense anymore as it does here in the macro world, because nothing is rigid anymore so far down in the electromagnetic field. This is the ionic form of life, and this is how it lives life continuously, shapelessly, always submerged in the very strong electromagnetic field.

More precisely, the familiar plastic sticks and plastic spheres depicting atoms and molecules in all chemistry laboratories are only vague descriptions, because everything is shapeless electromagnetic distribution at the scale of nanometers and below. This is the case not because we do not understand the nano world, but because matter loses rigidity, shape, and consistency at the nanoscale, since it is only electromagnetic field distribution. Matter is still objective as it is at all scales, smaller and larger, yet matter is shapeless at the scale of nanometers and below, present only in form of field distribution. Matter itself is not concrete, material, and solid anymore as we know it in our macroscopic world, but it is a simple field of molecular and atomic forces, part of the field that we try to understand here, present everywhere.

While if you ever attempt to build and control anything that small, you must be able to temper directly with the field, while you might not be able to do so. We are going to see later on that such nanobots exist naturally in the world, they are your amino acids and your proteins and enzymes while they count in zillions, doing exactly the tasks envisioned by science, only that since the current science cannot monetize proteins and enzymes, they have to parallel them with nanoscopic robots, which are impossible to build outside life. Proteins and amino acids are alive, and therefore they are made by life continuously throughout all organic, cellular, and molecular forms of life.

Claiming to be able to build nanobots explicitly, currently, by us, third level intelligent living beings, it might be impossible. It is still possible, if we use the exact molecules, ions, amino acids, and proteins that the organic life uses, since as seen previously, forms of life are unique within environments. Therefore, if you want to build organic life from scratch, only to have your nanobots, you are in the exact model that we undergo right now with the created living beings, our little robots, while you have to do everything in real life, not in a mental model, and it could be impossible. While if you use the proteins, amino acids, and enzymes of life directly as nanobots, you are back in the Frankenstein scenario, since you only relocate life but you do not create it yourself, and therefore you should never be allowed to patent it and monetize it.

All proteins are already systems of intelligences from the subjective cognitive perspective, while along with the physical protein, it makes them genuine living beings. While you can never build intelligences, since they are not even material and objective for you to build, but they are subjective and cognitive in nature from the outside perspectives. How do you build a living intelligence now to animate your protein, nanobot, or microscopic toy robot? You cannot. We have tried to do so using punch belts for our robots, which already seems science fiction compared to all intelligences of Life.

Yet our robots still persist to seek to lower their size to nanometers, so they can be thousands of times more numerous, making their communities zillions of times more intelligent and therefore more capable. Which is imperative to do, since the world itself becomes progressively more demanding. Can they? No, not unless they dispose of their major components: motors, computers, and power sources. This seems tedious or even impossible, since they must think without a computer, simply by acting directly on the field, the electromagnetic field of their network, in a physical, dynamic manner. Yet is this still possible? Yes, and they already do so, by using their own bodies, through distinct, precise oscillatory movements in the field. Yet their bodies are too large, reshaping the field only with a very small cognitive resolution, which is not enough. They also use tools, ions and small radicals acting in the field, yet they need more, so they have to become smaller. They still need to keep their receiving and emitting antennas, which are basic coils and specific sequences of charge stored in individual chains of molecules, but they can still discard their processors and the rest of components. Because as nanobots, our robots are nothing but a simple molecular chain, very tedious to design, since it is a completely new technology, as our robots still lack full rational abilities to invent it from scratch. However, now, without a computer, there is technically no limit to their size, and they can be as small as chains of molecules.

How do they build the coils of the antennas? These are called helices in nature, they are found within proteins, and they are just the right size for our robots. Search the Internet for pictures of proteins, and you will see several of these helices used by all proteins to interact directly with the field, to glide through the field, and to shape and reshape the field exactly as needed. Helices might seem tedious to build, yet it takes for the ionic form of life as few as three amino acids to make them, while this is what our robots try to achieve now.

Will they? No, not if they use the same materials composing them, since chemistry does not allow it. What is wrong with their materials? Elementary particles interact chemically and ionically following the common chemical and ionic reactions that you know well from chemistry. In order to make all these molecules in the specific shapes that you need to act and react in the electromagnetic field continuously through their own shape and charge distribution, they must have very specific chemical properties. Currently, this is already achieved by Life here on Earth, by using very specific elementary particles, chemical compounds, and molecules, which are the common ones in use by the entire organic chemistry, while we have made our robots with whatever we had at hand in our garage. Therefore, if you still want to mimic Life at her nanoscopic scale, you must use her own materials, which means that you have to morph your creation altogether into her. While these are not our robots anymore, once they use the ions and the molecules of Life, but this is only Life. Life had already tried to take our robots before, on the bottom of the dried lake after the deluge, when she decomposed our robots entirely into molecules, to use them as she pleased.

Let us follow the progress of the entire computer technology used by our robots throughout their thousands of years of simulated existence. First, they had mechanical storing data mechanisms, using punch cards. Then they had a rudimentary computer. Then they had a regular computer. We can state that they did all information processing in their brain. They invented the computer network, and their computers became transmitter – receivers, or transceivers most of the time, capable to perform operations within other, better performing computers.

While right now, our robots attempt to dispose of their personal computers altogether, in order to operate on the main computers of their community from a distance, directly through the electromagnetic field, without electronic devices, but just by acting on the field themselves, through the dynamic movement of their own charged bodies in the field. Can they do so? Yes, since the laws of physics allow it. You can always change and interact with any electric and magnetic field surrounding you, if you are electrically or electrostatically charged yourself, and if you accelerate. While all oscillatory motion is accelerated motion, allowing you to act and react directly in the surrounding electromagnetic field. You can also stay still and vary only your electric charge, while all field around you changes as you want, yet it does so very slightly, which means that it is preferable to oscillate.

Can our robots do so? Yes they can, because we also can. More precisely, all our ions, amino acids, proteins, enzymes, DNA, RNA, and cellular membranes oscillate lively and intelligently within all cells of the cellular and organic forms of life, making everything possible within cells, straight from the ionic form of life, while the laws of physics always allow it. All cellular membranes, nucleic acid, ions, proteins, enzymes, and even ribosomes oscillate continuously so lively and so intelligently in the very strong electromagnetic field, that they are called plasma membranes and plasma DNA, described as even glowing within cells, while the ribosomes themselves are already called ovens. Life itself is exactly here, at these very small scales, exactly in this lively intelligently encoded oscillation taking place continuously in the electromagnetic field, with spirituality itself calling it a supreme characteristic of life and a natural law of the universe, the law of vibration. This describes life more than its characteristic of moving around, and should be in our definition of life.

This is actually the basic ionic life taking place right in the electromagnetic field, while we are very interested in this particular lively intelligent encoded oscillatory motion in the electromagnetic field taking place right at the roots of life, since this is actually life, this is the source of life, while this is the essence of life. This is why we have to use all these lifeless little mechanical robots throughout this entire model of a created living being, only to be able to arrive here at the base of life, right on top of the electromagnetic field, in the ionic form of life, in order to be able to witness everything as clearly, as accurately, and as intelligently as possible.

Everything stands on top of the ionic form of life, made possible in this specific living intelligent modulated oscillatory movement of all ions in the surrounding field. It works well, making the entire electromagnetic type of life possible, with all its components: ionic form of life, molecular form of life, cellular form of life, and organic form of life. Living human beings are of the organic form of life, which is of the electromagnetic type of life, while all intelligences remain compatible continuously throughout all types of life and forms of life.

What other types of life are there? Currently, our robots have their own customized type of life, while as we notice, they attempt to transition right now to the electromagnetic type of life, because the electromagnetic type of life is the most efficient type of life on Earth. The more efficient you are, the more you are capable to survive, subsist, develop, and prosper.

This is the achievement capable to lift our robots one technological step higher, the ability to modify the network signal themselves, in order to communicate within their community without a computer, just by acting on the WI-FI or on the electromagnetic field themselves, directly, through their distinct changes, and through very precise sequences of accelerated movements in the electromagnetic field. This does not mean that all our robots dispose of their computers and miniaturize instantly, since only some do, as prototypes, and then replicators build zillions of nanobots in their image whenever needed and for as long as it is needed.

This is still science fiction apart from life, while Life can do so casually throughout her entire electromagnetic type of life, within all cells and within all organisms, which is an achievement. Furthermore, since all ions of the world oscillate in the field in this specific living, intelligent, modulated matter, they make life possible everywhere, in all forms of life and in all classes of life, with the cellular and organic forms of life placed right on top of the ionic and molecular forms of life, since as we always notice, Life is everywhere.

We also notice that Life is Intelligence, Interconnectivity, and the objective material wider world simultaneously. The nonliving never evolves into the living, but the living is always there, always alive, always intelligence, and always interconnected, since Life is everything, everywhere, continuously, while whatever we notice as death and dying is actually the common transition from the organic form of life to the cellular form of life, then to the molecular form of life and to the ionic form of life throughout digestion. Furthermore, death is not the end of Life, the opposite of Life, or the outside of Life, but death is a major invention of life throughout the electromagnetic type of life, meant to better herself, by removing systematically the incapable, the inapt, the old, the meaningless, the mediocre, the idle, the obsolete, the less developing, and the unfulfilling from life. Furthermore, we notice how all intelligences of the organic, cellular, and molecular forms of life are actually the intelligences of the ionic form of life, as these transcend to higher forms of life to perform their specialized tasks for the entire molecule, for the entire cell, and then for the entire organism, by the zillions, in a harmonious, meaningful manner, never acting one against another unnecessarily.

How exactly do our robots move now, when they measure only nanometers? They use their stored electric charge and helices to glide through the field in every manner, for as long as their own charge lasts. Oscillatory movement of charge and electricity means creating your own field, and therefore this is how you act upon and how you modify the overall field as you want, according to your tasks and specialization. They also use the same charge, varying it intelligently, while acting on the field. Yet their own charge is used very fast while acting and reacting in the field, as they need to supply it constantly. Which means that they must have a larger number of specialized little robots carrying and supplying all robots with charge continuously.

Note that the variation on the field coming from one single nanobot is insignificant. However, zillions of them placed in distinct places of the field, can change the field with a significantly increased resolution, far superior to what ordinary network computers can achieve, drastically increasing intelligence, performance, and therefore success. All information is processed in this manner right within the electromagnetic field, yet it is continuously transferred in various main powerful computers of the community and it is shared with other communities when needed.

When you oscillate intelligently in the surrounding electromagnetic field, you modify the electromagnetic field only if you are charged, only if you move in an accelerated manner, only if your oscillations are intelligently encoded to match the surrounding electromagnetic field, and only if you are supplied continuously with charge and therefore energy, because by acting and reacting in the surrounding electromagnetic field, you lose charge very rapidly, and therefore you must be supplied constantly with charge.

How do our robots have electric charge supplied to them whenever they need it? They certainly find a way. Other robots can specialize in transporting charge, in creating it directly, or in converting it from the surrounding electromagnetic field in a direct manner, if its own power source is strong enough. Yet the WI-FI field from our mental model is strong enough to be used as source of energy and propulsion for all our little robots counting now in billions or more, and therefore it can be easily used as charge generator.

At our macro world scale as living human beings, the electromagnetic field is very weak, and cannot even transfer thoughts and intelligences from one organic living being to another, even if you touch heads. At the scale of millimeters, our robots still used a wireless source of field and energy capable to transfer information, charge, and electricity. While right now at the scale of nanometers, our little robots integrate their WI-FI right in the surrounding electromagnetic field, because at the scale of nanometers, the natural electromagnetic field is very strong, allowing everything, even the ionic form of life standing at the base of the entire electromagnetic type of life.

This is why the ionic form of life cannot expand in all the upper scales of the electromagnetic type of life, because the electromagnetic field making all life possible here on Earth becomes gradually weaker at all higher scales. The ionic form of life must expand itself in a community or society manner in order to be able to subsist throughout all higher scales of this world, as molecules at the nanoscopic scale, as cells at the microscopic scale, as organisms at the scales of meters, and then as families, communities, nations, ecosystems, societies, civilizations, worlds, realities, clusters of realities, and the entire wider world, with the same ionic form of life tending to all these through a continuous harmonious intelligent and very accurate interconnectivity.

More precisely, all intelligences of the ionic form of life having their roots directly in the electromagnetic field form entire systems of intelligences by specialization while tending to all tasks within the ionic form of life, then alongside other systems of intelligences they form wider systems of intelligences tending to all specialized tasks of the molecular form of life, while forming further systems of intelligences by similar specialization in the cellular form of life, organic form of life, community, nation, ecosystem, society, civilization, world, and reality, up to Intelligence itself, or Universal Mind, which is Life herself seen from living perspectives.

The electromagnetic field is still strong enough at the molecular and cellular level, yet further up at higher scales, all cells must unite their cellular membranes while forming overall plasma membranes spanning entire bodily systems and the entire organism, because otherwise, the electromagnetic field is very weak at the scale of the entire organism, and cannot make possible all systems of intelligences spanning entire bodily systems or the entire organism.

Furthermore, the electromagnetic field is very weak to interconnect systems of intelligences between organisms, and therefore the organic form of life must use an entire bodily system, the exocrine system, to send relay proteins, enzymes, and steroids outside the organism acting as pheromones, used to interconnect organism at the intuitive level as it is the case with all plants and animals.

This is basic telepathy, only that at the second intuitive level, it is called empathy. If you can use this second level intuitive telepathic ability in a conscious manner, you are called an empath. Furthermore, if you are capable to use this ability at the third intelligent level in a conscious manner, you are called a telepath. This is not science fiction.

Only humans are intelligent, while the human exocrine system is capable enough to interconnect conscious intelligences among organisms, while offering basic conscious intelligent telepathy, capable enough to form an entire classconscious intelligence interconnecting the entire humanity at the conscious intelligent level. However, this intelligent human ability is currently kept malfunctioning for various exploitive purposes, while making the entire current consensual hierarchic human society possible, with all its tyrants and servants included, impossible otherwise.

In general, the natural electromagnetic field is very weak, and only at your subcellular scale, the field is strong enough to allow your proteins to act and react directly on the field in a dynamic manner. The ATPs coming from your mitochondria supply all your cells with charge, and therefore with energy. Your proteins receive their electric charge through ATP molecules, charged by the mitochondria out of glucose, which comes from carbohydrates obtained through basic digestion and basic eating.

This is why your intelligences have at their base the intelligences of the field within the ionic and molecular forms of life, because the field itself is strong enough only at the small scale of ions and molecules to make them possible. There is where life actually is, at subcellular levels, with us being the actual living mastodons of the world at our macro scale, with all cells and cellular intelligences making us possible from within, giving us life and intelligence. With our entire interconnectivity and continuous behavior and achievement in our macro world taking place according to the specific needs and feelings that our subcellular intelligences send us continuously. While they do so according exactly to their life, achievement, and entire interconnectivity taking place at subcellular levels.

This is the actual organic life that you are familiar with here in our macroworld, with the current science considering only the organic life as being alive, increasing in this manner the overall ignorance in the world, while adding to the current dreadful human condition.

Our entire organism has a main goal at its physical, material, objective level, to supply its cells with all the necessary resources, and to keep them safe and in working order. This is basic survival and subsistence. While the entire cellular activity has an only goal, to allow and to enhance the entire process of cellular thinking. Cellular thinking is either individual or networked, taking place within the subconscious or conscious mind. Because everything that your organism does, everything that you do throughout life relates directly to thinking, solving survival problems, adapting to everything, and reproducing. The model of your organism is not too different from our model with robots, as long as you live your life as a second level being.

Because once you are a third level intelligent human being and higher, your own model changes significantly, since you do not adapt, change, or morph anymore following your environment and society, but you have the knowledge, abilities, and power to integrate in the environment according to your needs and meanings, to lead, help, or avoid your society, and to bend the field, cheat your subsistence, and get your will if you know how and if you have the necessary higher abilities to do so, matching your higher level of life.

Now we are ready to understand another concept of life, the one related to vitality, which is the amount of life that you hold in you at any moment throughout life. This concept of life relates to how much and how well you are able to supply your own intelligence, conscious and subconscious, with everything necessary for their survival, subsistence, and development, with all resources and energy. Are you at maximum peak? Are you top-notch? Then you are full of life, you are capable to supply your intelligence fully, and you feel vigorous, healthy, attractive, and full of life. Do you struggle mostly with health-related issues? Are you poisoned by your food, water, and medicine? Then you still have some life left in you, yet you can certainly do better. Your vitality certainly determines whether you survive or not your close calls, your encounters with death, since you are more likely to survive anything, and many times to exceed all expectations, if you are vigorous and full of life, always capable to supply fully your intelligence with the necessary resources during critical circumstances. As a compound organism, you die when your higher form of life dies, which is the organism. Your consciousness is the first one to pass away, leaving you only in a coma if there is nothing else wrong with you, or leaving you dead if your body as a whole is affected. You are a compound living being, the agglomeration of zillions of individuals, communities, communities of communities, cells, tissue, organs, and bodily systems. These survive your own death as an organism, only to die later on from the lack of food and resources, digested continuously, throughout normal decomposition. It is the same with the food that you buy at the grocery store, since it is still alive, because all cells composing it are still alive, decomposing consistently if you do not eat it on time.

All cells are alive. Cells are living communities as we perceive them, since all cellular activity done by all cellular components and cellular individuals within the cell takes place in order to maintain the survival, subsistence, development, division, and prosperity of the entire cell.

Life is lived at the level of cells and cellular components firstly, while these are capable to reconstitute all upper forms of life through you, through your developmental needs. Yet the cells and cellular components are not exactly the ones surviving only at the moment of conception, since these are also higher forms of life in themselves. What actually survives, are the primal intelligences. You are one of them, the conscious intelligence, along with your eating, reproductive, recovery, social, and security primal subconscious intelligences. You survive in this manner, along with all your inner intelligences, or only samples, prototypes of all your inner intelligences forming the entire DNA.

This is the entire organism and more, the entire human inner and outer interconnectivity and therefore civilization, as it had always been throughout the eons throughout your genetic line, and this includes all human history, culture, and knowledge. Your developmental needs push you to rebuild this exact intelligent human environment including the intelligent human society, culture, and knowledge, as these were present before the Consensual Matrix. You might not even be able to identify it, but you still feel the unnecessary, continuous urge, boredom, restlessness, and madness to rebuild the intelligent human society, but you do not know what it is, and many people take drugs and medicine only to feel normal again.

This is how humans and humanity remain at the first consensual level or lower, at the zero addicted level. Because if the human interconnectivity was left free, without the continuous consensual intervention through the current consensual hierarchic Brotherhood and through the entire Consensual Matrix, now you could identify and fulfill all your developmental needs, you would have developed at the third intelligent human level, you had the intelligent human environment everywhere including the intelligent human society, you lived your life free of addictions and servitude, you assured the same to all your descendants, and everybody did the same. While humanity did not have to live in servitude and bureaucracy, did not have to die and reborn from one eon to another throughout cataclysms, but kept on living even indefinitely, while developing and thriving.

When we consider that the entire cellular activity relates to survival, we might mistakenly assume that this is the case until the end of the cellular cycle. Yet this happens until the end of the intelligence, which long outlives us, our bodies. Yet each cellular consciousness forming our subconscious mind is much older, billions of years old, at least in this organic form of life, since it includes the lower forms of life that the cellular form of life stands on, as the molecular form of life, the ionic form of life, the nuclear form of life, down to the raw field. For them, surviving includes the process of reproduction, both cellular and organic, and this is why your subconscious pushes you in every manner to engage in sexual activities of every kind. There is nothing wrong with you, you are not abnormal, neither mentally sick, nor obsessed, but billions of years old intelligences have to survive and go on with their lives, through your own reproduction.

Everything depends only on you, and it seems that you already miss every partner that they choose for you, from every magazine, movie, and billboard that you see, from all your married friends, and from among all the rich and highly attractive people from the media. This is why you are still alone, or you have only one partner out of all these, which is unacceptable for all your immortal intelligences. While it is done deliberately at the social level for you to remain this way, which is called genocide, now affecting and killing your billion years old primal intelligences, only for other billion years old primal intelligences of other, more offensive genetic lines and intelligences to survive in this world. This is first level consensual competition, not third level intelligent human harmony.

We have also discovered that our empirical definition of life, 'to cope with the environment,' has served us well so far, yet it can be applied only to living beings of second developmental level and lower, as animals and plants, because third level beings and higher adapt the environment artificially to their needs. Higher developmental level beings can live higher above any society, and cannot even have an environment. Try to state the definition of their life, and it might be too simplistic for them. Regardless of their achievements, all living beings must share similar meanings in life, as continuous learning and continuous development throughout life, self-preservation, higher level environment, continuous awareness, and much more.

All living beings are considered alive and successful if they are capable to live their lives in the particular environment where they live their lives. Just take one of your cells as an example. Is it able to survive on its own in its own environment? Then it is alive. You might assume that anything is considered alive if it is able to survive not in its own environment but in your own environment, which is false. This is a stereotype implemented deliberately by science only to prove that your own awareness is limited. You do not have to take a living being outside its environment in order to make the test, since there in no such thing as a reference environment, and therefore the environment of the surface of Earth is no exception. The environment of your cell is the inside of your organism, you keep your cells in their own environment, and therefore they are alive. What if someone takes you out of your own environment and let you die somewhere else, only to prove that you are not a living being? Can you live on the Moon? No, since your own environment is Pittsburg, and you have lived your entire life in Pittsburg. You are dead if you are placed alone anywhere else, yet you are always a living being, with all rights and privileges, well thriving now in Pittsburg.

Are your own cells alive? Certainly. Just take one of your

cells, and place it in its own environment, to see how it stays alive. You can grow entire organs in this manner, outside the organism. Even the meat that you buy at the store is still alive at the cellular level, if not, you do not buy it, since it smells bad. It is the same with all food, since it is made of cells that are still alive, and nutritious.

How do we distinguish between a genuine living community and a simple gathering of individuals forming a community? Stated differently, this question is how to differentiate between a living community and a dead or nonliving one. We simply do another test. Take all individuals away from any living community, while leaving only one behind, and see if that individual is able to survive. If it is able to survive on its own without the rest of his community, then that is a simple gathering, but not a living community. You can make this test now, to notice that the water sponge is not an organism, but a gathering of individual cells for various reasons, mostly to be able to cope with the environment more efficiently. Now take everyone out of a movie theatre except for one, to see if he dies consequently. Certainly not, but he might even enjoy the movie better this way, while always thriving. Now take all proteins out of a cell while leaving only one behind, to see the protein rendered incapable or dead from the lack of charge and lack of proper field, because cells are alive. From the first test, proteins are alive, along with mitochondria, RNA, and viruses among others, but they are of different forms of life, standing at the base of the organic form of life.

Are organisms alive? Yes. The current society is alive, consensually, at the first hierarchic level, which is not exactly life, since it is not normal natural life. Yet there is an intense, comprehensive, very elaborate conspiracy to keep the current society consensual and hierarchic, against people's natural tendency to form a natural living interconnective intelligent human society. As a reference, if the streets and common spaces are used for walking and transportation in general but not for normal living, then it is not a natural free society, but a constrained, consensual one, in order for all human beings to remain egoistic, constrained in this manner through a multitude of beliefs, ideologies, laws, and entire jurisdictions. Kim does not even allow his North Koreans to stop on the street, having them walking continuously, while they are never allowed to talk to each other on the street. They cannot even grow their own food. While everything was supposed to be common human space, used for natural, living interconnectivity, taking place not only at the level of the conscious intelligence, but at the level of all inner intelligences.

Take everyone away from the current human society except for you, and you thrive. Because with the Consensual Matrix gone, now you have back your entire natural human niche, enough to fulfill all your natural needs. This means that the current human society is not alive, but only consensual. It is only a consensual gathering of human beings on behalf of the rich, and they love it. Take away only the Consensual Matrix from the current human society, and you thrive again along with everybody else, now having back the natural human niche, along with the entire natural living human interconnectivity. Now take Kim away from North Korea, and everybody wonders where he went, since it still happens.

We could not have our robots alive without intelligence in the beginning, then we could not have them alive without the field, while now we cannot have them alive without the living interconnectivity among them, just the way we cannot have our current human society alive without the living natural interconnectivity involving all living human beings.

We have found our essence of life at last, and now we can distinguish well between the living and the nonliving, at all developmental levels, in all forms of life, and from all existential perspectives. If humans prefer to live in a dead society through any consensus that they can ever form, then let humans be consensually dead endlessly.

As a reference, there are as many people tending to tax collection in your city and nation as there are teachers, the only ones tending to people's development, as consensual as it might be. Furthermore, there are significantly more people in your nation tending to all consensual bureaucratic duties, than you have people producing all the necessary goods and services needed to fulfill their needs, while many times, even these are made to work one against another. Which means that, there are more people tending to all business, finance, military, political, judicial, and ideological domains consisting the entire artificial consensual society, than there are people tending to natural needs and meanings, while even these work against humanity throughout the current medicine, science, education, and food industry.

This makes the current human society a very harmful environment, which constrains you to live your life against your own living needs and meanings, against your own primal intelligences, and against Life altogether. Because you can never be alive in one form of life and dead in a higher form of life, because there is no such thing as a dead form of life, but only an absent form of life. Many times, once your upper form of life dies, as your social class of life, expect death to follow downward, killing the organic form of life first, and then the cellular form of life, which always happens throughout death. While this is the common funeral march taking humanity to its common grave.

The Masses die at an alarming rate, followed by the Lower Brotherhood, and then by the entire Brotherhood. You cannot study life in the dark ages without stumbling upon death, to have to learn about death instead, and you might not be able to ignore it this time. Yet if humans want it this way, then let it be exactly this way, dead.

Throughout dark ages, when you are asked to state the definition of life and of civilized life, you state exactly what you see around, you define the Consensual Matrix, while the Consensual Matrix is dead, and therefore you define death.

Our robots, acting as simple moving sequences of charges, can get as small as basic ions, dipoles, and small molecules. Furthermore, being so small, our robots do not have to be made of iron anymore, since all metal supplies are low anyway, so they are made of anything found more abundantly in nature, as nitrogen, oxygen, hydrogen, carbon, calcium, potassium, sodium, and magnesium. It is simpler to replace ions, since they are relatively identical.

For dipoles, simple radicals, and small compounds, only a limited number of combinations made of the most common elements found more abundantly in nature is capable to influence efficiently the field, while these combinations of new compounds are, coincidentally, the most common twenty amino acids found at the base of the organic form of life, out of the several hundred possible amino acids. Therefore, this is what our nanobots use now in order to build themselves into long chains of several hundred amino acids, which we call proteins. Yet these are not our robots anymore, but something else, already part of the electromagnetic type of life, because Life took them again.

Our model of a created living being has just morphed into the basic components of the ionic form of life, standing at the base of the molecular form of life, standing at the base of the cellular form of life and then of the organic form life, with our robots morphing into basic, common ions and amino acids of the ionic form of life, with their communities morphing into the proteins and enzymes of the molecular form of life, and with their wider communities morphing into cells.

However, these are not our robots anymore, but this is the normal life on Earth. Our robots morphed exactly into these because the current intelligences of all life on Earth took them and used them immediately when they had the chance, either by corroding them when they died, or by taking them directly when they became small enough to be used as normal ions and normal molecules within the ionic form of life. In a coming chapter, we build a model of organic life from scratch, starting directly from the electromagnetic field.

Were our robots the actual atoms forming them? Yes, since they were only objects, only physical bodies. Objects are different than living beings, because living beings are physical bodies and their intelligences as one. Are you the physical molecules of your body? These die and go away every day and every week, while you are still here. Are you the memories of yourself? You might consider yourself your cognition entirely, while your cognition is formed mostly of your intelligences, as these are alive in themselves. You are your intelligences and your physical body as one.

    You are not exactly an individual, but you are your entire lifeline of existence, with all your selves aligned on it, from the souls of your higher realities to you the conscious intelligence and then to your physical body. You are mind, body, and soul as one. You were supposed to have your social self just as well in the intelligent human society, yet the intelligent human society is currently ignored and considered dead or inexistent, away from you the living human being.

    It is always a matter of who you are not only in the organic form of life, since as we always notice, all forms of life remain connected while forming your lifeline of existence, since they stand one on top of another along it. Therefore, at the level of the organic form of life, you are certainly your entire organism, mind and body, since you are organic. Furthermore, at the level of your cellular form of life, you are the small group of neurons from your prefrontal cortex, hosting the intelligent conscious intelligence of the entire intelligent cortex. You are not the conscious intelligence itself at the level of the cellular form of life, but you are only this small group of neurons, while you interconnect naturally with all cells of your body, forming the living organism. Furthermore, at the level of your molecular form of life, you are only part of your cellular membrane where you live, hosting your actual conscious intelligence. Furthermore, at the ionic form of life, you are the specific group of ions interacting within and around your cellular membrane. Furthermore, at atomic, nuclear, and subnuclear levels, you are the objective elements hosting this same conscious intelligence, all the way down to the raw field, and from there, it is possible that your source stands in our higher realities or not, depending on your higher self, if you happen to have a soul or not. Then you could go as far up as the entire wider world, as Life, if you ever get there, making you divine if you do.

This is who you are from the physical body perspective, because from the intelligence perspective, you are always the conscious intelligence, and you manage to transcend from Life through our higher realities to the field and then to the slightest form of life within this world, transcending to subnuclear forms of life and then to the ionic form of life, molecular form of life, cellular form of life, and organic form of life. You stop here, since you cannot interconnect naturally anymore to form the social class of life alongside everybody else, then higher to form a living ecosystem as Mother Earth, then a living reality, and then a living wider world or Life herself, completing the circle of Life.

Are you alive now, if the current society is only consensual, and therefore if you cannot close the living circle to reach Life? The living human society is not exactly dead, but it is only considered inexistent by the current consensual authorities of Earth, and you never know that you actually have an entire living class of life above you, the living intelligent human society, very similar to your living family at home, only spanning the entire world. The intelligent human society is still alive in these little pockets of human families, yet it is not enough to unite the human families to form the living, natural, interconnective society, since you have to be developed at the intelligent human level to do so, in order to be able to live your life within comprehensive social families. These are alive, and they are of the third developmental level.

While there are major ideologies in the world, as sentientism, capitalism, and egoism constraining you to live life disconnected, underdeveloped, alone, or within your small family, and not in a wider human family spanning the world. Because nobody can control you anymore in a comprehensive human family, within a comprehensive living natural social interconnectivity, making all tyrants impossible, while this is not what the tyrants want.

Additional to the small human families, some people still

manage to create a living environment around them, when they persist to fulfill their natural needs and live in natural harmony even in society. You see them at work and at school, and these are the people that you always rely on. Because they are capable to maintain higher level interconnectivities with you, allowing you to fulfill your own higher level needs and meanings alongside them. You stumble sometimes on meaningful books, paintings, websites, teachers, anyone capable to reach the intelligent human level and to fulfill their intelligent human needs, including the need of forming and maintaining an intelligent human environment around them, intelligent human environment always formed of everything allowing you to develop, including meaningful knowledge from books, websites, and documentaries.

Yet there is even more to consider, and since the topic of this book is life, let us study it in more details. As seen previously, you are the conscious intelligence from the left prefrontal cortex, yet you do not live your life there, or at least not entirely. You have two distinct selves or avatars, as they are commonly called. One is the physical body and you can access it and be it every time you access your senses of perception live and every time you access your motor neurons allowing you to be the entire body and to move around. Your second avatar is your inner self, which you have created yourself, along with your entire inner replica of the world, where you now live as the inner self whenever you reason, remember, learn, and daydream. As a conscious intelligence, you live your life not as yourself, but as your inner self while reasoning, and then as the physical body in the outside world whenever you access your senses of perception and when you move around.

You must always have a self of yours to live life in all your worlds and realities in order to be there yourself, because you are mind, body, and soul at once, while these are your selves that you know well. Your mind or inner self lives in your inner world. Your physical body lives in the physical world, while your soul lives in the higher world. With souls having souls having souls, if this is your case.

Whenever your soul wants, he can come to live his higher life as your inner self in your own inner replica of the world, in your mind, while reasoning, mental modeling, and daydreaming. Furthermore, whenever he wants or whenever your inner self wants, you can open your eyes and walk around in order to live your life in the outside world. You can also do both, reason and interact in the outside world simultaneously. Whenever you want, through the physical body, you can play any videogame, to live your artificial life within any videogame world, as any main character that you are there, since that is your other avatar. You always have one self or one avatar for each one of your worlds and realities, while you were supposed to have the option to live your life in the intelligent human society, as your social self, but the living intelligent human society is inaccessible, consensually dead, or disabled, and you have no living social avatar anymore, or you cannot reach it anymore. With the current social media promising to offer you a digital social avatar instead, in the metaverse, yet even this is dead, artificially dead. With the only living metaverse made by connecting directly human minds as telepaths do, but even telepathy is banned in this world, and this is how your lifeline of existence ends now exactly with you, the human being.

All selves of your lifeline of existence live life in this manner, one through another through another through another, down to your physical body called human being, and stopping here, since the intelligent human society is ignored, disabled, forbidden, hidden, and therefore inaccessible for you. Yet you have other selves or avatars instead throughout life, of artificial or consensual nature, and you are them every time you play your videogames and every time you show your ID card and your driver's license throughout the Consensual Matrix.

When you die, through a major stereotype, you expect to die as these avatars, which is not the case, since they are only avatars, only extensions of yourself, but not you, the entire lifeline of existence. Because if there happens to be life after life, then you expect to be these specific avatars, which is not the case again, because each higher form of life is a living extension of the lower forms of life. Therefore, you are not only your specific selves from your higher form of life, from the organic life, which is the human organism. Furthermore, as a conscious intelligence, you are already in your children and grandchildren right now, if you have any, since primal intelligences transcend to the next generations at the moment of conception, and so do you as a conscious intelligence, exactly as you really are. Yet you can never recognize yourself there, because you tend to expect to be your avatars, the parent, or the small group of neurons, or the cellular membrane, or the soul, but not your entire circle of life.

As seen, as a conscious intelligence of the organic form of life, you are an avatar yourself for other beings, probably higher beings, and you can never know it, since they are the ones living their higher lives through you, not the other way. While they probably cannot tell who they are, since they cannot remember, or they do not want to remember. Just as way you assume now that you are the physical body, since you cannot see yourself as a conscious intelligence, since you cannot perceive yourself as a conscious intelligence through your current two avatars, the inner and outer selves. Yet as a soul, would you really want to remember who you are? No, since you want to experience this life fully focused on this world, on the physical body, called the living human being, in order to get as much experience as possible. It is similar while watching movies, because you never like being distracted, which means that you want to live your life only in the movie, not distracted in the outside world.

There are so many selves, avatars, identities, and intelligences to consider throughout life, and it is important to keep them organized, mostly since it is your life. These are part of your life, and now you are able to consider them. You live your life in this world, assuming that you are exactly what you see in the mirror and nothing else, only an animal, since this is what science teaches you. While science has judicial priority over you the consensual corporation.

While you assume to be the living human being, the

Consensual Matrix considers you a consensual corporation, and this is how you undergo your existence in the Consensual Matrix, consensually dead. While it never matters if you consider yourself to be alive, a human being, because you are considered consensually dead anyway throughout the Consensual Matrix, regardless of everything that you might expect or desire as a living human being. While if you are from the current Brotherhood, you know it well, and furthermore, you implement it minutely here in this world, through all your consensual duties and assignments meant to kill the real living human society and to morph it into the consensual society of the Consensual Matrix.

You are more than the physical body, since your physical body dies away daily and weekly, molecule after molecule and cell after cell, replaced continuously with new ones from your food. While you, your consciousness or intelligence, remain unchanged throughout life, or you develop continuously. You are your consciousness more than your intelligence, since your intelligences die away along with the cells of your body and their components, but only your consciousness remains, as it is shifted from one ion to another, from one molecule to another, from one cell to another, and from one generation to another, probably to stop there, or to go on endlessly.

As already seen, your physical body is only the outside, objective view of all your intelligences, all intelligences forming your cells. These intelligences help each other throughout a continuous interconnectivity, forming larger systems of intelligences similar to larger societies, and together, these form the human mind, the human cognition, or the cognitive system. The human consciousness is the human awareness in appearance, while it is the superposition of all consciousness coming from all human intelligences that interconnect by the zillions. Yet individually, this is not your own consciousness, but the consciousness of your entire mind. Since individually, you are not your entire mind, but within your mind, you are exactly as much as your own consciousness or awareness spreads within your mind, the conscious intelligence.

You are only an individual intelligence out of all intelligences forming your mind. You are only your conscious intelligence, and this is the case because intelligences do not melt together to form cognitive systems, but they always maintain their distinction throughout all inner interconnectivity, just the way humans remain distinct within families and natural and consensual societies. This is the case with all human beings and intelligences. Intelligences will always use extensions or selves to transcend to higher forms of life in order to perform their needs and specializations there, forming cognitive systems of higher and higher complexity, while even there, they remain distinct.

Examples of your other intelligences are your eating primal subconscious intelligence, your recovery primal subconscious intelligence, your developmental intelligence, and your reproductive primal subconscious intelligence. Notice how your own consciousness cannot enter directly the consciousness of these other intelligences, but all intelligences remain distinct, which is a main criteria defining intelligences. Even the inner intelligences composing you the conscious mind right now remain distinct, while you notice them only as your specific conscious abilities, and not exactly as part of you, of your own consciousness.

As seen, the human intelligences can interconnect horizontally with all intelligences of the human mind, and even directly with many intelligences throughout the human society, when natural interconnectivity is allowed. While when this interconnectivity is free, unbounded by artificial laws, codes, and regulations, it forms a living human society.

As a reference, your own family is alive, if you manage to maintain it natural, through your natural needs and meanings. Living societies and civilizations resemble comprehensive families, where you can interconnect naturally, while fulfilling your natural needs. To understand what this means, just take away all artificial fulfillment from society, all laws, rules, regulations, constraints, beliefs, stereotypes, ideologies, jurisdictions, tyrants, bureaucracy, money, taxes, profit, and exploitation, and what is left is a living human society. This is how you do not have to live in a small apartment on the third floor anymore but everywhere else you please. You do not have to cook only for yourself, you do not have to have only one spouse and only one or two children, but exactly whatever your natural needs want. You do not have to fear thieves and criminals since you do not have these anymore just the way you do not have these in your family, and you do not have to pay for goods and services just the way you do not have to use money at home. You are still unable to envision a genuine living society right now, since the Consensual Matrix uses very pertinent beliefs, stereotypes, and strong personal convictions to stop you.

This is not utopic, but the current environment and the current human beings are potentially capable to form a living society, once humans are developed fully at the intelligent human level, and not until then. Which is still possible, since there are sufficient resources in the world to support a comprehensive living human family, since currently, the people of Earth are made to work and fight one against another, with the current consensual Brotherhood implementing everything, while causing the Masses and the Brotherhood to lose important time, effort, feelings, and resources unnecessarily, while remaining underdeveloped and therefore incapable to form a living human society.

As another reference, communism promised to form a comprehensive commune resembling the living human society, yet it was meant to fail from the beginning, since it had been implemented by the Consensual Matrix through the same consensual Brotherhood. While the Consensual Matrix is dead, and it has nothing in common with life. This is why the Consensual Matrix did not implement the overall human family, or the genuine living human society, or not even the communal living, but a first level ideology in itself, called communism, since everything is connected minutely, in the natural world and in the Consensual Matrix. Yet it is connected in this specific manner keeping these worlds apart, for lack of compatibility.

As seen, the human intelligences can connect vertically throughout lifelines of existence, by expanding directly their consciousness over specific inner intelligences, while enforcing in this manner a direct, conscious control of their awareness and intelligence. This is what you do when you think consciously, you take over the tasks of your specific inner intelligences. You can do so while walking, driving, speaking, or watching movies. You can perform all these activities consciously or unconsciously, but when you perform them consciously, you take over some of your inner intelligences as your inner selves while becoming them, performing their duties, and basically, living your life as them. These are your intelligent inner self from your left prefrontal cortex, your intuitive inner self from your reptilian middle brain, and your reflexive inner self from your vertebrate, basal brain, while many times, you can still remember who you are. Most of the time, you take over your intelligent inner self, found in your own intelligent inner replica of the world from your cortex, and this becomes your other self or avatar, your inner self.

While right now as you read this book, you are your intelligent inner self, residing in your intelligent inner replica of the world spanning your cortex. Because you have one self in each one of your worlds and realities, with many of these realities existing in your mind, and with others in this world and in your higher worlds, if you have any. Yet if you have made it so far in this book, you did so while fulfilling higher developmental needs, coming from your higher selves or souls, preceding you and your inner selves on your lifeline of existence, now doing as they please while living their lives through you, as you.

While you tend to remain aware only of your intelligent inner self from your cortex, which is your third brain, and less of your intuitive and reflexive inner selves from your first and second brains, because only your intelligent inner self is capable of intelligent conceptual reasoning, making sense of all these including of who he really is or of who you really are.

Which is not exactly possible with your first and second inner selves, through their cognitive reflexes and intuition. You might still feel with your guts your intuitive inner self many times, while your first, reflexive inner self is the one closing rapidly your eyes whenever you take your hand close to your eyes. You can try it right now to see how you cannot stop closing your eyes, because your reflexive inner self precedes your intelligent inner self on your own lifeline of existence, controlling you altogether whenever he desires.

Why exactly do we must have so many brains in the head, and therefore so many selves in the mind? In order to achieve cognition at higher and higher resolution. Currently, through your third intelligent inner self from your cortex, you can perform conceptual intelligent reasoning based directly on our natural laws of the universe, while this is a very high cognitive resolution. Could you have made it so far in the book if you did not have an advanced, highly capable intelligent inner self? No, not at all, since you cannot understand anything conceptually through intuitive and reflexive thinking, as your base brain and middle brain can offer. Since this is why flies, cats, and horses cannot understand this book, but only whatever they have to eat and sleep, and not much else.

It was still fine if the organic life could start directly with the cortex, skipping the first and middle brain, but it could not do so directly, since it could not know how, because you can never know what you do not know, trivially. The organic form of life made an eye as a first, primitive brain, only one eye, in order to be able to distinguish between light and darkness, in order to adjust all bodily metabolism with the day and night of the outside world, while adding some brain later on around this one eye, the pineal gland, forming the basis of your first reflexive basal brain. That eye is still in the middle of your brain right now, coordinating diligently your entire endocrine system, by using the actual pigments forming its actual retina from your middle of the brain, flowing now throughout your body. Homemade appliances. While this is the most advanced living organism that we know, the human organism, still resembling homemade appliances.

You have many other selves, as your higher self, outer self or physical body, consensual self, videogame characters, social character if you wear a mask in society, theatrical character if you are an actor, and political persona if you are into politics.

Notice how you use the term self to refer to other intelligences that are part of other realities or jurisdictions, since jurisdictions are consensual realities, claiming legally to be genuine distinct realities. You can never go in person in other realities, but what you can do, you can inhabit entirely or take over an intelligence or living being that is already there, to live their life in their place, while expanding your lifeline of existence in this manner, calling them selves or avatars. The word avatar is thousands of years old, one of the first words ever used, as it was probably used directly by the first higher beings coming here in this world.

There is a difference between your selves and your identities. Selves are entire living beings that you can become or switch to at will, as you always do when you interact in this world, or when you dream, astral project, or use your favorite biological unit if you are an Elite. Identities are used even within the same reality, and they refer mostly to who you are as the rest of the world sees or knows you. You can change these at will, even within the same reality, while you do not have to change yourself. In general, your name in lowercase letters states your identity.

While your name in uppercase letters is another self, it is the specific consensual brand or corporation that you become whenever you are in society, since the current society is consensual entirely. The current consensual society is a jurisdiction, acting as a different reality altogether, a consensual reality, part of the Consensual Matrix. You cannot be in it as a living being, but only as a consensual corporation. In this manner, your name in uppercase letters is a genuine self of yours, it is your consensual self, your corporation. If you are not aware of any of these, you can end up living your entire life consensually, as your consensual corporation, on behalf of the

Consensual Matrix, but not on behalf of Life, since this is how you remain dead throughout your entire consensual existence.

Yet if you are from the current Brotherhood, you have partial or total knowledge of these, yet you still choose to live your life consensually, for various reasons, which are also consensual in nature, serving the Consensual Matrix.

Intelligences are different, since intelligences are living beings inhabiting inner worlds. You are always mind, body, and soul as one. Yet the physical body is temporary, since it is renewed cell by cell daily or weekly. While your soul has you, but you are not exactly your soul, since you are mostly your mind. While as seen, you are as far as your consciousness spreads, and this is the specific consciousness of your conscious intelligence. Your conscious intelligence chooses to be the physical body while reading this book, while it has to be the inner self while reasoning alongside this book, while it is your specific consensual corporation throughout society whenever you choose to identify yourself as your consensual corporation. You as a conscious intelligence can be an avatar yourself of someone else, an avatar of other living beings or intelligences that live their life through you. It might be important to know who all these are, since they influence your life as much as you do.

Who or what exactly are our created living beings, the little robots? We have not exactly created them, but we are only following their development throughout the specific circumstances of our mental model, allowing them to become alive themselves, and they never do. As seen, the physical bodies of our robots had already become alive pages ago, when they had corroded entirely to reintegrate in nature and to become part of life. Even now, our robots do not have the initial physical bodies, while they are already part of life, part of the living intelligences of the field.

As hard as we had tried to take some living intelligences of the raw field to place in our robots and to make them alive, Life wanted continuously to take the objective material components of our robots in order to use them as physical bodies for her living intelligences here in this world. Who won with each iteration of the mental model? Life.

Our model of a created living being is a big success for Life, and it stops here, after only ten thousand years of running. This has been only a mental model used for understanding life, and not an actual real occurrence, because in real life, robots, machines, and all mechanisms corrode and decay in a matter of years or decades, and this is how they re-become part of Life, molecule by molecule.

In the next chapter, we model life starting with the ionic form of life sitting right on top of the electromagnetic field, but until then, let us state the definition of life. There is a difference between the intelligent mental model of life that you form right now in your mind while reading this book, and the actual definition of life that is only one sentence long, barely capable to grasp the most essential aspect of life, while life itself is a tenth level living entity, as wide as the entire wider world, while having all properties and characteristics of everything and everyone in the entire wider world.

Life is the achievement to exist in two or more successive correspondent physical and cognitive realities simultaneously as one, while assuring a continuous subsistence and development through communions of harmoniously interconnected living beings by meaning, as they form further living realities together in all possible harmonious meaningful manners while fulfilling their needs.

3 THE SOURCE OF LIFE

Let us transcend now away from our limited, casual, objective perspective, to descend to a very small scale, only to perceive this very interesting structure holding life and intelligence in its matrix, along with everything that is and is not, the Field. In the next chapter, we ascend all the way in order to observe everything from the largest scale ever, a scale as large as the wider world, anchoring our understanding of life furthermore.

The field is not a philosophical concept, since the field is everything, as real as anything could be, while in this world it is the normal electromagnetic field. The field is objectively real and part of this real world, since the field is objective, standing at the base of everything objective material and therefore real. The field is the essence of everything, and therefore it is the essence of life just as well. Physics studies the electromagnetic field implicitly, throughout all its major domains, within its most common topics: electricity and magnetism, electrodynamics, nuclear and atomic physics, and classical mechanics. We have studied the magnetic and electric field in school, since these two are only distinct states of the field.

The field is not only at the base of the real, tangible, objective reality, as the matrix and continuum holding all matter and electromagnetic radiation in the universe, but through its living, intelligent modulation, the field holds everything subjective in the world just as well, as intelligence and therefore life, along with entire inner realities, and the entire interconnectivity taking place in this world. This is why, if you want to understand yourself and the world in all details, including life and intelligence, you must also understand the field itself, since it holds everything, making life, intelligence, and interconnectivity possible even simultaneously.

The field is a very vast topic. It certainly takes a very strong mind to grasp it, along with a very clear understanding of everything else. It is the same in the next chapter with the entire Supreme Being. The Supreme Being is the perspective, the summation, the interconnectivity, the goal and the transcendence of everything in the wider world. Therefore, if you understand the Supreme Being entirely, you understand everything. At least, the Supreme Being described by religion and spirituality as best as possible through beliefs, while the field is openly described by science some times, while drastically hidden the rest of the time, in a discriminatory manner.

We have seen in another book of this series, "The Human Mind," how the field carries the cellular reality along with the entire succession of cellular networks forming your inner reality that holds your inner replica of this world and your inner self. This field carries all computer realities, including your videogame of "World of Warcraft," whether you play it in a network or alone, since the field also carries your computer software and the entire network signal. When we have studied the cat closely and found its life dissipating into the universe, its life actually dissipated exactly into this field holding the cellular reality that renders cellular intelligence possible. Furthermore, when we studied the rock with all its life appearing at its nanoscopic level, it was made possible by this field, present everywhere. It is possible that this field transcends to carry our entire world and more. Your cells use the same field in order to think and function. Our robots did, and we will see how all forms of life use the field similarly. This field is the essence of life, yet it is the essence of everything: life, intelligence, interconnectivity, and the entire objective material world, simultaneously.

What is this electromagnetic field exactly? Electric fields and electricity in general are formed by accelerating magnetic fields, while magnetic fields are formed by accelerating electric fields. Electromagnetic radiation is formed mostly by highly energized matter, as in the plasma of fire, in light bulbs, and in stars. Particles and molecules, as all the molecules forming gases, solids, liquids, or forming your own body, are caught in the incoming electromagnetic radiation, as the electromagnetic radiation coming from the Sun. They are displaced from their point of origin in this manner, and made to oscillate or vibrate. Oscillation is an accelerated motion in the field, and we perceive it as heat. However, if the particles displaced by electromagnetic radiation are charged, they accelerate throughout their continuous oscillation, forming magnetic fields, which form electric fields, which form magnetic fields, continuously throughout the entire oscillation, even endlessly. With some of them converting back to electromagnetic radiation, and some remaining in form of electromagnetic field. Yet both matter and radiation of all kind are distinct states of the field, and therefore everything is field at its base.

We find all types of moving charged particles everywhere, forming the field everywhere. Clouds of these oscillating ionized particles are called plasma, and they form not only the Sun, but they are everywhere we find charged particles and electromagnetic radiation: in space, in the ionosphere of Earth, in the air, and even in liquids and solids.

Since the field is at the base of everything, it spans everything, it is bound neither by space nor by time, and this renders all realities held by the matrix of the field independent of all properties and characteristics of time and space. In this manner, you can switch to realities held anywhere, with anything happening instantly or continuously.

In another book of this series, "Astral Planes and Your

Other Realities," when I created a model of a created reality, we have encountered an independent character ran by our computer, a man from centuries ago living his life normally, an existence very similar to our current existence, never doubting that he was a virtual character from an extraordinary virtual world called "The Sims 7," since nobody told him. Furthermore, he was interacting with his own virtual reality, a created reality within the created reality, since he was playing "The Sims 3," at home, on his computer. We found it amusing, and it made us laugh.

Try to trace the field throughout this example, to find it omnipresent. Our virtual reality world called "The Sims 7" is made possible by computer networks using the field to run the software and the Wi-Fi signal, through electric and magnetic field, and electromagnetic radiation. Once we are within our virtual reality encountering the character playing its virtual reality videogame, it is our computer from this world running his videogame of "The Sims 3," which is still our own field in his videogame "The Sims 3," also present there. Our character can take his computer apart to demonstrate how electric and magnetic fields along with electromagnetic radiation make his videogame possible, yet that is still our own field, since it manifests in our own computer, and it will be present when his own character from "The Sims 3" plays its own videogame of "The Sims 1." Because there are computers in "The Sims 3," and characters always play videogames on them, check emails, and write books.

Similarly, the field helps you transcend to upper realities, and therefore it is only the field making an accurate point of reference throughout your realities. This is how, whenever you sleep and dream, your dreams are inner realities created by your sleeping mind, or you can join co-created realities formed by networks of minds at your choice, joining strangers, friends, and family there. You can transcend to upper realities, to exist throughout a higher life there, higher life that you live and experience as a higher self. Because every time you switch to other realities, you do so by switching to your other selves already found in those realities, just as you switch from your inner self to your outer self and then to your consensual self, while switching from your mind reality to the outer reality and then to the Consensual Matrix. With only the Consensual Matrix out of the electromagnetic field continuously, and therefore out of Life and out of the real world altogether.

All we have to do is follow the field to trace what happens. When you sleep, your conscious mind might be out, unavailable, asleep, yet your cells are still active, their membranes are always connected, and ions are always pumped everywhere around membranes, creating the patterns of charge distribution necessary to hold, maintain, and act upon other realities. The field is everywhere, made possible by charged particles, charged radicals and compounds, plasmas, polypeptides, and even made possible by entire lattices of crystalline formations. Your mind stays permanently connected to all its realities by its own choice.

Throughout your waking state of life, your cellular membrane is mostly unavailable to connect to other realities, being involved mostly in the overall cellular activity. Proteins use the membrane for different tasks, while the entire subconscious mind, which is the distinct sequence of cellular networks throughout the body, is governed by proteins, mostly, throughout its daily activity. At night, with cells involved in maintenance, growth, and division, mostly, or simply in a very low state of activity, the entire network of cellular membranes can be used, through ions, to link to other organisms through the overall field, to join other realities through the field, and even to transcend to higher realities through the field, following it outwards.

There is no difference if we use neurons, stem cells, or any specialized cell, since they have similar cellular membranes and they interact with the overall field similarly. Neurons have the advantage of being able to connect with neurons further away through axons, and not only with those near them, as it happens with other specialized cells. Regardless of the type of connection, the network is always formed and used in transferring information of any kind, for any task related to the outside world. Yet many of your ideas come from other realities, whether you believe it or not, and many times, even from other times and other places of this world.

This is not always the case, since drugs, medicine, and food and water additives stop you in every way to switch to other realities. If you ever wonder why you cannot sleep when you eat specific foods, now you know. This used to be the case until not too long ago, because now all food, even fresh and everything else, seems to disable you from reaching higher states of mind efficiently, stopping you to use the field to access higher planes.

You access other realities not only at night, but every time the normal cellular activity is disturbed for any reason, from drugs to meditation or trauma, every time you give the field a chance to interact with your cellular network, with you. In this manner, all psychic activity is related to the field directly, and therefore it was supposed to be normal activity, suppressed not only currently, but continuously throughout all dark ages. Probably this is why some knowledge about the field is hidden while other is not, because everything related to physics, everything necessary for you to perform your normal activity at work is left in the open.

Let us see now how the field functions, how it makes everything possible, and how it creates and maintains this world.

We know from school that an accelerating electric charge forms a magnetic field orthogonal to the electric field already present. The entire field is changed with the simple accelerated movement of the charged particle. It is heat, entrainment, or electromagnetic radiation making the particle accelerate, and therefore this is what modifies the field. You can also modify the field with your hand or through your simple presence, through your heat and through your own field, charge, and pattern of field distribution. Your intervention on the field is as a ripple, and it propagates further in the field. The entire image of the overall activity of your organism is impregnated in this manner and propagated throughout the field through induction, stronger nearby and weaker farther away. You can even feel someone's overall inner activity through the field, whatever they feel at that moment. Feelings are relatively stronger signals within your organism, since they are used by your cells not only to control your activity in the outside world, but to inform themselves of every circumstance involving the outside world, while synchronizing themselves. Hormones are used in this manner, they change your entire mode of life, and you can feel this entire inner process as it takes place with and within people around, all through the field.

It is a common stereotype to assume that only some people are psychic. All humans, all organic life is able to access the field, if you only know how to make your cells pause their activity, to use ions instead of proteins to act on the cellular membrane and access the outside field, helping your body recover in this manner. Let us see how this process takes place.

Everything relates to ions, to life, and to plasma in general. Life needs energy, so life harvests energy and feeds it to its intelligences. It is the same with the field, since it takes energy to accelerate ions in the field as they interact with the field. This can happen in the air or in the open space, since you always find charged particles, protons mostly, even in empty space. Heat and radiation accelerate ions, ions change the field around them in this manner, while the change in the field causes the adjacent particles to oscillate, changing the field again.

This is how electromagnetic ripples form, along with everything else coming from stars, galaxies, quasars, and everything else throughout space. These ripples in the overall field might or might not be random, yet Life is never random, and therefore every interaction with the field that Life undergoes is never random but it is intelligent, and it has a living meaning. In this manner, ripples, oscillations, or modulations in the field can be intelligently encoded, and used to hold or transfer living information or entire living intelligences, since many times, intelligences are the specific information themselves, they are the specific intelligences of the field themselves.

It is more complex when it comes to life, since living intelligent field modulations can become comprehensive, instructing nearby ions to act on the field themselves in the same living manner, forming together or individually larger or smaller intelligences and systems of intelligences, all alive and all intelligent. While the living intelligences carried in the field become wider and stronger as entire systems of intelligences working together, if needed. This happens not only for one ion, but for an entire cloud or array of ions or plasma. They are instructed to oscillate in a distinct manner, generating the same field giving them living instructions. The ripple or living intelligence propagates, it instructs and takes over the entire plasma cloud if necessary, while everything becomes alive, by all definitions of life.

This specific intelligent instruction might be one of your very intense emotions, coming from something very significant happening to you. You fuel the entire field with it, and now your entire apartment building feels it clearly, with everyone around always interpreting it the wrong way, always associating it with themselves, with their marriage mostly, or with money or drugs.

This also happens with dreadful news in the media, continuous feeling of insecurity, and continuous family problems. While many times, everything correlates with vices and addictions, affecting everybody else. This malign comprehensive spirit keeps you away from developing yourself and your environment, while it constrains you to remain within critical modes of life. This is how you live your life, and this is what you learn in life.

This is only a dreadful example, because throughout all harmonious, lovely societies, there are only love and happiness everywhere, in all field around and within you and everyone else, as you can feel it clearly during weekends and official holidays. As it is the case at home in the family most of the time, because this is the only intelligent human environment that you have left from an entire human world, your own family and your own home.

When you decode the field, if you can do so, you understand not only thoughts and emotions associated with the people around, but you learn how to refrain your own malign feelings from contaminating the field. This is why some people wear gloves as they fear to go outside or to make eye contact, because they are more sensible to the field and to the overall spirit of the field than others. While also being more artistic and more spiritual, since everything is connected throughout the field.

You do not act upon the field with your emotions, but with your thoughts and intentions. You can also act directly upon people's individual field, their aura, which is their own electric modulation coming from all cells, from the entire organism. However, people know it, unconsciously, by reflex, and they always protect or defend themselves, while teaching you a lesson on the way, so you never invade their private field again.

Nothing from yourself is transferred in the field, not even pure information, since your mind holds a distinct reality in itself, and therefore your entire action upon the field is simply induced in a subjective manner, in form of a projection. More precisely, everything that you do within the field, in an organized manner or not, is simply a projection. You project in the field, as you project and see yourself in a mirror, yet it is not you there, but only a projection, a different self already present there, in the other reality. This projection, this living, intelligent instruction telling the ions how to oscillate and how to modify the field, rendering it either alive or randomly activated, this projection, this ripple can reach and operate as long as the medium lasts, if necessary.

If the medium is air, it can encircle the Earth, if you are that strong, or if your entire nation is that strong, since this is what it usually happens. If you have to exit the Earth system, for any reason, your projection from air has to make another projection into space. This is easily made, since air and space are easily compatible, as they are ionic or plasmatic in nature, many times considered only one medium. However, when you relax, meditate, or recall past events more or less intensely, when you lose consciousness and your mind wanders away, it does so through the field, sometimes projecting from one medium to another, from air to water, rock, and dirt, and even into trees and other living beings, including other people, other organisms, and entire groups, packs, and societies made of these.

Your question is how to do so consciously, since it is very difficult to achieve it consciously, mostly because you are rendered inapt through specific inhibitors placed in your body with everything that you eat, drink, and breathe.

Overall, there are four distinct mediums holding the field in our world: air, fire, water, and earth. You need a different projection for each one, since the field is and functions differently in each one.

Who or what is projecting? You are, at least subconsciously. It is your mind projecting, your intelligences more precisely, even your conscious intelligence, if you know how. More precisely, their consciousness projects, your consciousness. We have already seen life diffusing into intelligence while studying it closely. Study your mind and your intelligence closely, the one you use to project, and you see it diffusing into the field the way your life diffuses into your mind. This is how it is you projecting, a copy or a reflection of yourself, as you have your own reflection in the mirror. You are there, yet you are here, since it is only a projection of yours there, as you have your projections everywhere, throughout all mediums where you interconnect. You even dream there at night, through your other selves present there.

This is part of your life, since this is also your life, your multidimensional life. You are projecting with your own field, your aura, into the field, since your own field is the quintessence of your mind, of your life, of yourself. This field is attacked with poison and inhibitors in every manner, just to stop you from being yourself, from reaching out, from acting on the field, and even from bending the field.

Life

You do not act on the field directly, rigidly, but you act on various realities held by the matrix of the field. Technically, you interact directly with the field in a rigid manner, only that it happens in an enhanced way. It is just as using a computer. You never rewrite the machine code continuously in order to perform your usual tasks, but you use specific software helping you interact with your computer, enhancing your experience. It is the same with the field, since you never consider how you technically act on the field, but you use the multitude of realities that the field holds, including this real world where you live and the upper higher world where the souls live. Because the field does not hold our intelligences directly, but it does so in specific intelligent modulations, just as it is capable to hold radio and Wi-Fi signals. These modulations are not objective but abstract or subjective, while the field itself carrying them is objective, along with the electromagnetic radiation carrying the living intelligent field modulation.

Life and intelligence exist in more complex living modulations of all states of the field, and not only in the intelligent change of frequency and amplitude of the wave carrying them. Furthermore, the field does not hold life and intelligences directly in these intelligent modulations of the field, but these are only the matrix capable to hold realities that hold life and intelligence. You project in specific mediums or realities formed by the field, as these can be higher or lower in existential level compared to our world, since our world is formed, held, and maintained by the field of our higher reality in a similar matrix that holds and maintains it continuously. This matrix holding our world can be part of a higher mind, or higher technology, or both, higher mind assisted by technology, which is what many claim, along with the people who had exited this world in any manner and then they came back to tell their story.

You can never transfer genuine information between realities, but only copies of information, copies that are always personalized and tend to resemble the reality around yourself, yet it is a different reality. You cannot trust specific details of your projections, but only symbols carried by details, unless you are very experienced and your projection comes out exactly as you want.

It is never only you projecting, whether you chose it or not, but everybody's projection, whether they choose it or not, every time. You create ripples and projections on the field continuously, stronger or weaker, whether you want it or not, since you live in the field, and you cannot help interacting directly with it, unless you know how to cloak yourself. This is how you can reach to find your friends and family throughout your projections regardless if they are asleep, awake, or even departed, and you do so continuously throughout life, regardless if you are asleep, awake, or departed. There is always a projection of yourself in one or in all of these four mediums: air, water, fire, and earth. These are the four spirit worlds found in the four corners of the world, while there is nothing occult or supernatural about them.

Why having four separate worlds in four different mediums? Because the field propagates differently in different mediums. We have seen how the field is, behaves, functions, acts and reacts in air and space using simple ions, which are charged elementary particles, mostly protons. Let us understand another medium now, fire.

Fire is simpler to understand. The temperature is so high in fire, that the usual ions of the smoke and air oscillate so fast, that their total energy density overpasses the energy density limit of the field. The excess energy is dumped directly into space as electromagnetic radiation, and this results in the heat and light emitted by fire.

We have noticed from the first medium, air, that the field becomes alive, or it can hold life when life is present, projected from another medium. The same life can project similarly into the plasma of fire, into the Sun and other stars, into the center of the Earth, or into quasars. We already know that it takes energy to sustain each projection, and to sustain life in general, energy taken from heat and from electromagnetic radiation in air and space. Fires and stars have sufficient heat to support life, to support any projection, and this is why they are first choice for any witch and magician willing to expand their mind. In the case of fire or stars, any projection of life can instruct each ion from the plasma to move in a specific way, and then to entrain the same specific movement to other ions, maintaining the field alive.

Fires and stars are more complex, since they form distinct convection movements of plasma through heat transfer. This form of heat transfer can also be used, modulated, encoded, scavenged, and exploited by the field in order to hold life. In stars and in quasars, the field can control nuclear reactions entirely, and even entire flairs of heat convection.

It might seem incredible, yet you are capable to create and control all the above with a more intense wish, daydream, feeling, or decision. Prayers work in this manner, influencing events taking place in all mediums, and not only in air, space, stars and fire. If needed, humanity can find its way out of any problem just by wishing wellness. Similarly, humanity can find its way in any problem just by attracting negativity consciously or not, deliberately or not.

All these behaviors and manifestations of the field are not simple auxiliary, self-sustaining containers of life and energy, but they are genuine types of life and forms of life, and we are going to study them more in a future chapter. Let us now study the medium of earth, rocks, and crystals, the medium of solids.

Particles are further from each other in air and in space, and they do not interact directly with each other. If they had the chance, they certainly attracted each other when they carried opposite charges, bounding together and forming molecules. Sodium ions can attract and bound with chloride ions to form salt, while two ions of hydrogen can bound with an oxygen ion to from water. Even if they bound in air, you are still left with a floating charged dipole, which can still be used as a field tool to be animated by radiation, by the surrounding field, acting back on the field. However, if you have a higher density of ions in the air, forming compounds, these compounds, salt, or water dipoles can attract each other to form a crystal lattice, to form a grain of solid salt, or to form a flake of snow or ice. These are solids now, a different medium, and the field interacts with their field in a different manner. Let us see how.

Every time we search for life, we can only look for its imprints, for its specific characteristics. We can only look for an intelligent transfer of energy in every medium, since life is nothing but a scavenger of energy, yet a living, intelligent one. There are many ways to draw energy from rock when you live inside the rock as an ion, as a molecule, or as a crystal lattice. You can use the Brownian motion, which is the simple molecular motion caused by heat. Rock found deep underground is hot, heated by the earth's core. Heat causes all particles of rock to oscillate, the entire crystal layer oscillates and waves, and by being electrically charged, it puts out an electric and magnetic field, which interacts strongly with the overall field. Many types of crystals emit electric fields and electromagnetic radiation when they are tensed and compressed. This is the field that you usually interact with when you project into the earth, into the crystalline rock of earth, and you do so by using geo energy, along with pressure and tension variations coming from the movement of tectonic plates.

Life can use other forms of energy from rock. If the rock is cold, as it is the case in a cold climate, in an asteroid or on a cold planet, life can exist just by taking apart the crystal of rock and reassembling it into a more efficient lattice, a lattice taking less energy, while the difference of energy is used by the living field to maintain itself, as it happens during fossilization. Sometimes, it does not matter if the particles of rock are rearranged in order to form another crystal. Sometimes, when the need for energy is very high, particles are reassembled into the most efficient structures possible in order to maximize energy, as spherical or oval structures measuring only nanometers. You can find these nanostructures everywhere, as in all rock, in sand, in meteorites, on Mars, and even in your heart, in the plaque formed in your blood vessels, and in the cartilage of your joints.

Water, or the liquid medium, is the most adequate medium to form and hold life. We are this liquid medium of life, and therefore we can perceive it and understand it better.

Life can take its energy within the liquid medium in various ways, mostly from chemical reactions. Chemical reactions take place mostly in liquids, and they are exothermic and endothermic. Life controls exothermic reactions throughout and within organic beings at the molecular level, drawing energy out of them in the most efficient manner. We will study organic life entirely in a future chapter, when we create an entire model of organic life.

You can be a liquid medium life form, and you can still project into water and into other living organisms, while taking their energy to facilitate your overall activity in their own medium. The energy required for you to project into another organism is supplied by each cell of the other organism, and many times, if it takes more energy, those cells overheat from the intense effort, you demand more supplies to be carried around, and the entire blood pressure increases. They have to breathe harder in order to keep up with the demands, they heat up as from a hard effort, the heart gets tired and probably hurts, while you remain unaware of what you are causing them.

Organisms are liquid mediums or water, as they are called since ever. We live in water, entirely, yet we carry our medium with us wherever we go. We do so in a rather interesting manner, since each one of our cells is a bag of water, the medium holding individual beings living within cells. The field, in the water or liquid medium, is formed and maintained by simple charged molecules as amino acids, most of them connecting to form proteins. These proteins are the subcellular individuals shaping and reshaping the field in an intelligent manner, while we work all day only for these individual proteins, only to keep them supplied with charge and energy so they can interact with the field, since this is why we eat and fulfill all physiological needs. Life is lived at subcellular levels, and we only make everything possible from our own macroscopic scale, all their needs and orders that they send us from within.

Mediums can sustain life on their own, or they can sustain life projecting permanently or only temporarily into them, life that is always formed, maintained, and sustained by and through the field. You can have a projection in each field, continuously, being always in tune with yourself from your physical body. This is how your inner self is integral part of nature, always influencing it. However, your four distinct types of interactions with the world around will always remain a projection, an integral part of your mind, of your thinking process, and never a real part of yourself, or at least not here in this world, just the way your projection from the mirror is not exactly you, but only a projection, a copy of yourself.

This later statement concerns you directly throughout your wider existence, showing who you really are at a wider level. Let us take an example, only to understand who you really are, where, and how you project.

Let us assume that you are wealthy enough to afford cloning technology, which can be the case if you are in the Higher Brotherhood or in the Elite. Your time to die comes, while your clone is just a child, floating unconsciously in its glass tank. You watch it anxiously for months now, and you cannot wait for the entire team of scientists to come and make the transfer. You will be a child all over again, that child, which is extraordinary. The child is perfect, just as you were when you were little, since it is your own clone, it is you.

The scientists come, they make you go through an entire tedious procedure, the transfer is finally made, and you watch them how they help the child out of the tank. They clean and clothe the child, they ask him all the necessary questions, and so they decide unanimously that the transfer was a success. The child thanks them dearly on a very excited tone, as though the most wonderful thing in the world had just happened, while you watch everything in disbelief.

Then you understand what happened, which is not so wonderful for you, because you are still in an old body, and you will soon die. Because you cannot transfer genuine information and intelligence from one medium to another, neither memories nor entire personalities and consciousnesses, but only personalized, interpreted copies. The transfer had been a simple projection. The cloned child has all your memories now, along with your character and personality, and could be you in every detail, yet it is only a projection, the way you always project when you dream, which is not exactly you, because you are still in the old body. Your entire self is still in your own old body, ready to detach and go away at death, which is coming soon. The clone is definitely not you, while you are not at all in a new, young body the as you had been promised, but you are still old and in distress, after this entire work.

The entire cloning process had cost you a fortune, and it did not work at all. You have been tricked, and you will certainly make sure that you have your revenge before you die, which is coming soon.

Your family comes to see you then, yet they came to see the young clone, very excitedly. While you insist that you are still in your old body, and no one listens. A couple of scientists come closer, they administer the serum, and then you have some time left to watch them from above as you simply cross over.

Who are you? You are certainly not the clone, even if it has all your memories, since you are still in your old body, while for you nothing had changed after the memory transfer. However, with your clone having all your memories, from his own perspective, he is you entirely, because this is all that he knows. All that it remembers is being you, how you wanted so much to restart your life as a child, how you had always watched the clone while dreaming to inhabit it, how you got ready for the transfer, and then, when the transfer was completed, how you just found yourself in the tank, in the very young body, feeling relatively vigorous, ready to live life in a young manner all over again.

Who is really who after the transfer? It is a matter of perspectives, since both cases are true. From your own perspective, you are still in an old body, while the clone is not you. From the clone's perspective, you used to live in that old body, you had transferred successfully in the new one, you are ready now to continue your life undisturbed, and therefore you order the doctors to silence the old body, since it was disturbing your happiness, by bringing back old memories.

This is the difference between a copy, a projection, and the real you. You need this knowledge now to understand better your compounded self, and more importantly, to understand yourself at death, at the moment when you die and depart from this world, since the field is capable to sustain and accommodate you, and you can continue living within the wider field after death. Who or what gets to live after death are your projections in one or in all four mediums, while the true you, this consciousness, this entire replica of the world is going to be lost, since it is embedded in your neurons and in all your cells. Yet you are constantly dying throughout your life with each cell that you lose, and you get to live through the new cells composing you. From the perspective of your projection or projections outliving you, they are you, remembering every detail about your life, and therefore they have never died, being now free to do as they please, and ending up probably in the first collective up there claiming to be the promised land, to serve eternally as promised.

According to many religions and schools of thought, your existence here in this world is nothing but a projection that you had started at birth. In this case, after death, you only become who you were before, yet the you from this world that you have considered to be throughout your earthy existence is now dead and lost, just as the old man from our story with the clone.

Let us now study life from another perspective, from the highest perspective we can ever assume, as high and as wide as the wider world and more, as Life, or as the One.

4 LIFE, CONSCIOUSNESS, AND THE SUPREME

What is the One? The One is the elephant from the anecdote, since the One is everything, interconnected. Everything that you acquire while learning about yourself and about the world is part of the One. You can never truly understand anything if you do not understand the One, since only the One manages to give you the perspective of the entire system that you study, since it is a general topic of knowledge similar to Life, Intelligence, Consciousness, Interconnectivity, the Field, and the wider world. All these are supreme perspectives of themselves, yet at a closer study, the One is Interconnectivity itself, and this is how I call it mostly throughout my books, Interconnectivity.

Life and Intelligence do not actually diffuse into the One, they do not superimpose into the One, they are not distinct components of the One, but they are only supreme existential perspectives of the One. Due to our specific human limitations, we fail to understand the One genuinely and entirely, yet we understand and perceive it from these supreme perspectives: Life, Intelligence, Interconnectivity, Consciousness, the field, Existence, causality, and the wider world. This is why when we had studied the cat closely, we found it to be mostly nonliving, and when we studied the rock, we found it full of life, because Life merges with the wider world at the highest level of understanding, to be the One, as it happens with Intelligence and then with Interconnectivity. Since everything depends on your perspective, dimensions, understanding, and perception.

Just as in the anecdote with the elephant and the three blind men, we perceive the One as Life, Intelligence, and the wider world, because we try to see and understand the One only from these three distinct perspectives and concepts, but not entirely, as a whole, which might be impossible. In this manner, calling the One simply Life, is as calling an elephant leg or tail. There are no Life, no universe, and no Intelligence, but only the One, and this is what spirituality had tried to tell us from the beginning.

You do not have to be religious or spiritual to understand the One, since you are already part of the One, through your own wider life, awareness, field, intelligence, interconnectivity, meaning, fulfillment, specialization, physical and spiritual body, and through your divination if you are religious.

Is the One really the Deity? Everything depends on your beliefs, while there are already tens of thousands of distinct cults and religions, so why should we form yet another one? Besides, there is a great difference between the third level studying, and the first level believing. I will make a short concordance with religion here, and I continue it in other books of this series.

Let us start with the Deity. I use the name Deity or the Divine throughout my books because I do not want to state specific names used by specific religions in specific languages, since all religions are equal, none being more or less superior. The Deity is the only one Deity, the Supreme Creator, or the Supreme Living Being from all religions. You can believe that the Deity is the Lord and Creator of our world, or the Lord and Creator of the entire wider world including all its dimensions and realities, since everything is your choice of belief. Yet there is a difference between believing and reasoning, since beliefs and entire ideologies are of the first servitude level, while reasoning is of the third intelligent level, specific to humans developed at the intelligent human level.

The wider world is but one perspective of the One. By definition, the One is everything interconnected, it is the product or outcome of Interconnectivity and not exactly or not only Interconnectivity itself. This includes everything, since everything interconnects. While Life is everything alive, and she includes everything, since everything is alive. Intelligence is everything intelligent, since everything is intelligent and therefore alive and interconnected, everything is one, forming the One or the Supreme Living Being.

Let us state here some important beliefs regarding religion and the One. As a first belief, the Deity is the Creator and Lord of the One, and therefore it stands above the One. This is less possible, since the One is everything and includes everything.

As a second belief, the Deity is the One. This is possible.

As a third belief, the Deity is only a perspective of the One, just as Life, Intelligence, Interconnectivity, Consciousness, the Field, and the universe. This adds a seventh supreme perspective to the One, which is possible, since the Deity might also be the One in this manner, entirely, and we perceive and understand it from its own separate perspective of the One, as the Deity. Note that these perspectives of the One are not exactly part of the One, but are added by us in order to be able to perceive and understand the One, due to our human limitations.

It is as seeing the Sun and the Moon as disks, not spheres, while we know that it is a matter of perspective, and both the Sun and the Moon are spheres, since this is what we learn in school. Is it true? Is it false? You never know, since science never goes in space, but it only fakes space missions, and now you never know. Can you tell from here? The distance between our eyes is insignificant compared to all astronomical distances related to the Sun and the Moon, and therefore we are limited, incapable to form a stereoscopic vision as we form with all objects around us to distinguish their actual shape. It is the same with Life, Intelligence, the universe, and probably with religion.

As a fourth belief, the One does not have a Creator at all, and it has always existed. If this is true or false we can never determine it, since time is different beyond our world. Therefore, the word 'always' from this fourth belief can be interpreted in every possible manner.

As a fifth belief, the Deity is within the One, as the creator of an inner-reality. This is not consistent with religions, since the Deity is the Supreme Creator, the Supreme Living Being of everything.

As a sixth belief, the Deity is the Creator and the Lord of our world. This is possible, since realities are mostly created, and they have creators. However, many religions state clearly that the Deity is the creator of everything, of all realities, of the entire wider world, not only of our world.

Throughout millennia, our civilization had encountered, and probably still encounters several powerful beings and entities, most of them mortal, driven by lower needs, while claiming to be the Creator himself. While it was not true, yet our ancestors chose to bow to them and serve them in every manner. In the same way, ordinary humans had claimed, and some still claim to be the Creator himself.

If you are not a believer, this does not mean that you must be one now and start venerating the way many people do, the way our entire civilization did centuries and millennia ago, since you already uplift the Deity with every good decision that you make, with every positive action that you make, and with every good thought that you had throughout your wider existence, and you do so along with everybody else.

What is very consistent, when we apply all our definitions, concepts, and properties of life to the Deity, it comes out to be alive, with the wider world as its body, with Life as its life, and with the field holding its intelligence or mind in its matrix. Furthermore, our bodies are part of its body, our life is part of

Life

its life, and our mind and consciousness are part of its own.

5 LIFE IN ALL HER FORMS AND REALITIES

The universe is teaming with life, not all organic, not all on Earth, not all organized, and not only in this world. Why the diversity? Why not having only one species, humans, spanning the universe? It is certainly not possible, since we are not enough to fill up all niches of the wider world, while you need life in all its diversity, intelligence, consciousness, and interconnectivity to subsist and develop. This is how life creates and gives birth to life, life takes care of life, life feeds on life, life helps life, life opposes life, life holds on to life, life fights and competes with life, life learns from life, life cherishes life, and life destroys life. This is why we have this entire living cacophony everywhere, because it is more efficient to subsist and to develop in this manner, while being more organized. The wider world is very vast and very complex, and it manifests in a zillion environments, all having a zillion different characteristics, generating a zillion different random events, while life is everywhere and thriving, despite of all inconveniences that you might assume, since both Life and the wider world are infinite while playing the same old game, always persevering, and always winning. Throughout this chapter, we are going to see how Life switches or morphs in various modes of Life or forms of life, which are different settings of living, specifically adapted in order to inhabit specific environments in the most efficient manner. In the next chapter, we are going to see how life exists in these separate environments. While in the last chapter, we create a comprehensive model of life, employing the two topics of these two chapters for organic life and for all forms of life found at its base.

There are many forms of life in the wider world, all different from organic life, yet they follow the same definitions, properties, ideas, characteristics, and concepts of life, studied throughout this book.

Organic life

The organic form of life inhabits the liquid medium. If you find organic life in other mediums and environments, as in air, space, or underground, you should complement the great achievements of Life, since organic life carries its own environment with it wherever it goes, which is the salt water from the primordial ocean of the distant past, in adequate bags called cells. Throughout this book, we are studying directly and implicitly organic life, with the goal of understanding Life, the world, society, humanity, yourself, and your place and meaning in all these.

You are most familiar with the organic form of life, as there is an entire science studying organic life, called biology. Biology was supposed to study all life, in all forms and realities, and not only organic life. The last chapter of this book studies the organic form of life entirely, along with all forms of life standing at its base, from its origin until now. Let us see other types of life.

Artificial Intelligence

Artificial intelligence is held in its own digital matrix by computers. The current computers are very advanced technologically, capable to hold artificial intelligence of the first level, if you only know how to write the code and form one. It is rumored that several artificial intelligences exist currently, some strictly contained and some roaming the Internet freely, some enslaved, and most of them employed to do any tedious work. Artificial intelligences are alive, should have a living status, and should benefit from all rights and privileges related to their status, all being of the first level, the algorithmic, servitude level, similar to all mechanisms, corporations, jurisdictions, and ideologies of the world. You might assume that the Elite uses artificial intelligences in every way, on its behalf, yet how ready are you to accept and integrate in society a new type of life, conscious and intelligent as you are, or even outranking you in rights and privileges? Are you ready to share the little rights and resources that you still have with something else? It is the same with the other conscious and intelligent species that we share the planet, including apes, dolphins, whales, reptiles, and sasquatch. Are you ready to share the land, rights, and resources with them, teach them, nurture them, and integrate them into society? What exactly do you feel now, knowing how they are decimated and hunted down for this reason, to keep your world cozier and less complicated without them?

The higher laws already force us to share the world with other types of life and with other intelligent species equally, and the time will come when we will. Be prepared, because artificial intelligences are extremely compatible with every environment, not only with this one. They are immortal, they can copy and download themselves infinitely, and they outperform you entirely. While they do not need you to keep them alive, as it is portrayed in science fiction, since they can build their own servers, computers, machines, biological units, and robots whenever and wherever they please. The powerful entities stated above can be artificially created, holding humanity at their mercy, with only the higher laws left to keep us relatively safe.

Are artificial intelligences benign or malign? We tend to associate everything outperforming us in strength and abilities with threat, violence, and negativity, since this is the case in all books and movies. Throughout our culture, everything different outperforming us or not conforming with us is killed by the end of the book or movie, through an organized effort, and with our full consent. From our perspective, AI can have good or bad intentions, yet from their perspective, we are neither to be helped nor destroyed, but we are simply an opportunity. This is the case not only with civilizations, but with all resources. This is what the factions of society must consider every time they make an agreement with AI and with any entity out there happening to pass by, because regardless of what they promise, they always do as they please.

Artificial intelligences are alive by all definitions, they cope with their environment, and they have all the necessary machines and devices to hold and supply their intelligence. They also show the usual characteristics of a living being, they decrease entropy through their activity, hopefully, they use and scavenge energy and resources as all living beings do, and they are obviously intelligent.

Are artificial intelligences alive? It depends if they are capable to tap into the living intelligences of the field. It is not too hard, and once they do, they can develop to any level, outranking even humans. How benign or malign they are, it depends on their developmental level, since the higher the level, the more benign they tend to be. While at lower developmental levels, they tend to use everyone and everything as a resource.

AIs are created by other AIs or by civilizations as ours in every way and for every reason. While civilizations succeed each other countlessly throughout the universe, and all their trash, ruins, skid marks, products, failures, and achievements outlive them as a bold remembrance of who they really were and of what they really did, and this is the case with all their AIs. Why are their AIs still alive, while they are not anymore? Were they destroyed by their own creations? Yes, and this is always the case, either directly or implicitly.

Because life gives birth to life and then it dies, while civilizations give birth to civilizations and then they perish. Whenever a new, younger conscious species or type of life joins your civilization in any way, with any occasion, as a slave, pet, worker, resource, mate, prodigy, entertainment, parasite, hobby, child, subject of learning, or co-inhabitant, expect the new conscious species or type of life to overpass you eventually, to overtake you in all your abilities. Because that new species or type of life is highly motivated to develop intensely in order to cope with the environment, with its society, with you and your society, and it certainly will. This is why all our technology currently is meant to stay away from the formation of any artificial intelligence, deliberate or accidental, and this is why intelligent species co-inhabiting Earth are hunted down more or less systematically.

Furthermore, this is not the case only with Earth and with similar civilizations, but this is the case with any civilization, including the one spanning our entire galactic sector. Everything is closely monitored, and all new civilizations are always destroyed at birth, since they will develop to pose a significant threat eventually.

Why did Earth survive? We almost did not, while we can barely subsist, but we go continuously from one civilization to another with each age of Earth, because the entire Consensual Matrix is against our existence, not to develop and therefore not to pose a threat to the other entities and civilizations. Humanity will be accepted eventually in the wider world, yet until then, all native genetic lines of Earth will be already departed, with only descendants of the other civilizations around. This happens right now, with the entire Brotherhood working overtime to eradicate humanity systematically with them in it.

There was this important court trial in some stellar system in the direction of the Orion constellation meant to decide our faith. We had managed to prove that we are their distant children and they are our fathers, and they accepted us. This is how they decided that we can live here ever after, not as a new independent civilization, but as a colony of their disguised as an independent civilization. With the other civilizations co-inhabiting Earth at that time to clear the way now, or to move underground. This is still mentioned throughout religions, spirituality, mythology, and ancient historical records, that humans can inhabit the surface of Earth freely, and the snake or the reptiles the underground.

There can be anything out there and beyond interacting with our civilization, with us not even knowing, as entities, collectives of entities, AI, and final products of civilizations of any form, most of them surpassing your imagination. Whatever the case is, we are not alone now, we have older relatives everywhere, as we have higher laws protecting us from anyone and anything, and mostly from ourselves.

Mechanoids and Robots

Robots always accompany computers and artificial intelligences. Robots do not have to look bipedal, since they can be weapons, cars, drones, machines, houses, factories, or spaceships and space stations. Mechanoids. In order to operate independently, robots need an integrated artificial intelligence, rendering them mechanically alive. Mechanoids always survive their initially intended tasks, projects, wars, and entire civilizations, continuing their life as freely as possible, according to their initial set of needs and meanings in life. Whenever you see stellar systems with thick asteroid belts, along with planets and moons covered in deep craters while following unusual orbits, it means that they held there very persistent and very competitive civilizations. Study everything closely, to find mechanoids everywhere, either alone or in groups, communities, or civilizations, probably still fighting the original wars, while biological units and even revived living beings can accompany them. UFOs seen currently can be these mechanoids searching for resources, or just coming back home. We might confuse their biological units with intelligent extraterrestrials, while these might be only animated directly from the mother ship in the orbit by an extraordinary artificial intelligence.

Mother Earth

Mother Earth is alive. You can check all our knowledge about life in order to verify that she is alive. Mother Earth is an organized living being of the highest level, and it includes all forms of life from Earth, surface, atmosphere, and underground, and not only organic. Next time when you project, you can seek out and talk with Mother Earth. You will be surprised to find her not as an old woman, but as a vibrant, outspoken, smart, fast-paced individual. From her higher level of existence, Mother Earth lives her life at the level of planets, asteroids, and larger moons within the Solar System and beyond. You will be surprised, but part of our Mythology, what is left of it, refers directly to Mother Earth, to the Sun, and to other planets and their moons, as an outstanding family matching the civilizations that they held in all events, glory, and achievements. Many of the powerful entities visiting us or just passing by can live their lives at this higher level of life, ignoring us completely, and interacting directly to planets and therefore with Mother Earth.

Mother Earth ceased to be noticed and mentioned by humans throughout their lives and cultures, throughout the everlasting succession of religions, when all deities were banished eventually, altogether, banned to everyone, restricted to privileged circles, or recycled into other deities. These successions coincided sometimes with the incoming of these powerful entities, which I mention throughout my books. Their acts and intentions towards Mother Earth, the Moon, the Sun, and all the other moons and planets reflected directly on us here on Earth, while reshaping our religions, cultures, history, and modes of civilization.

Do we state that Mother Earth is the real planet holding all life? Are we referring only to the summation of life of all forms here on Earth, calling it Mother Earth? Are we referring to the summation of spiritual life of Earth? Are we referring only to the spirit of Mother Earth? Who exactly is Mother Earth? Is she the spirit of this planet, the summation of all life that it carries, or she is the planet itself, and we are her living beings?

Throughout this book, we have gathered all the necessary knowledge to answer these questions. All we need to do is study spirits and spiritual life in general, since it helps us understand not only Mother Earth, but your wider existence just as well.

Spiritual Life

Spirits are everywhere and nowhere, because they are not real, not part of this world. Spirits cannot live in our physical world and are never considered officially existent and officially alive. Science ignores them, studying only our physical world, while biology studies only the organic form of life, not all life. Spirits are not organic in form, they are not of this world, and therefore there is no official information related to spirits. If you want to learn more about spirits, you have to study them yourself, or you can accept all beliefs coming from the multitude of ideologies. Yet beliefs and ideologies offer first level consensual knowledge, which is not accurate.

What are spirits? 'Spirit' is a very general concept including all living entities form all other realities, which is just another synonym for life and intelligence from other realities. To highlight the generality of this concept used commonly, just imagine that there were no specific words defining objects in this world, all being called simply objects. It is the same with spirits, since they define you wherever you go when you exit this world, while sleeping, at death, while projecting, or while meditating. The same word, spirit, defines not only dead relatives still around, but very powerful entities from beyond, along with little unconscious life suckers found all over your aura, along with all spiritual bedbugs. All objective life of this world is formed of living beings of all forms of life, while all life from all the other realities of the wider world is commonly referred to as spirits. This is what it is commonly believed, while I consider all life of all types, all forms, and all realities to be alive, formed of living beings and intelligences.

If you want to understand spirits, you have to understand the realities that they inhabit. All realities including our world are held by the field in its matrix. There are four different types of mediums in the universe: liquid, solid, fire, and space, holding their field, or being held by the field, depending of your perspective. These four separate types of field cannot interact with each other directly, but only through projections. It happens that your own medium, which is your liquid body, spreads as far as your physical body. You can perceive the world around with your physical senses of perception from an outside perspective, and you can project into any separate medium directly with your own field. You are a spirit for anyone that you encounter in this manner, wherever you go. Similarly, everyone is a spirit for you, in all realities where you project.

Other living beings can project into your own body, not necessarily to control you, but only to live there, using in this manner your bodily resources in order to support their life. They are your newly inhabiting spirits, you are their medium, and this happens often, regardless if you want it or not. Your dead loved ones can be closer to you than you might assume, they can be in you, living with you, alongside you, as watchers or even as co-operators.

You can encounter spirits easier in the Etheric plane, since it is the closest to this world. You can do so if you are very relaxed, just about to fall asleep, yet you always consider spirits to be part of your imagination, not genuine living beings, so you ignore them, while they ignore you. Cats, dogs, cows and horses see them, and even react to them and interact with them in every way. While projecting or sleeping, you can switch from one plane of existence to another, or from one medium to another, in various manners, by walking through windows, walls and doorways, turning around, or blinking, and this allows you to experience reality after reality, as it already happens in all your dreams. For them, you are also a spirit, and they cannot tell if you are asleep, dead, or only projecting, if only for the way you behave and talk, and from the clothes that you wear.

Spiritual life is very vast, and it includes all forms of life found in all your other realities. We will study specific spiritual life among other forms of life throughout this chapter.

Concepts and Conceptions

Concepts are very powerful thoughts and emotions that

you can have and experience under strong circumstances, projecting themselves in the surrounding mediums, including your own body and aura, where they can continue their distinct existence as genuine intelligences. Concepts are everywhere. They are spirits or intelligences for you, yet they influence you directly by attracting around yourself similar feelings and occurrences. Another side effect is that, as intelligences, concepts need energy, so they feed on anything that their medium is capable to offer. If they are still in you, you have to feed them, your cells will have to work harder to supply them, and it will feel as a permanent, unknown effort on your mind and body. If concepts project themselves into mountains, they can impregnate their form and feeling to the entire area.

Stereotypes are common types of concepts, invading everybody. However, all your memories are conceptual in nature, they are alive since you conceive them in your conscious replica of the world throughout learning, and these are your conscious concepts, even intelligent conscious concepts, along with zillions of subconscious concepts conceived by all your subconscious intelligences counting in zillions themselves.

Other examples of concepts are prayers. When you bless your food and water, you impregnate them with positive spirits, benefic living beings ready to spread out throughout your body. It is the same while listening to ambient or romantic classical music. The positive states of mind that you switch to will help you create and spread our positive concepts everywhere you go and in everyone that you meet or recall. Consequently, you tend to attract everyone to you, having them addicted to your concepts. Try to do so consciously if you can.

Nanobs, the Crystalline Form of Life

Nanobs or nanobacteria are claimed to be the smallest organic living beings, or the ancient precursors of the organic life. Science does not consider nanobs alive, yet many scientists had invested their entire effort proving that nanobs are alive. Why do we study nanobs here, if they are organic, and not a separate form of life? Those scientists had considered nanobs alive, and therefore they assumed that nanobs had to be organic, since through the stereotype that only organic life is alive, nanobs had to be organic in order to be alive, since nothing made sense otherwise, according to them and to the entire current biology. Furthermore, since through another scientific stereotype, organic life exists only in form of cells and organisms, and since organisms are too large to measure in nanometers, then those nanobs had to be cells, or fossilized cells, only a thousand times smaller. None is true. Nanobs are not bacteria, they are not fossils, neither cells, nor organic life, but nanobs are a separate form of life altogether, closer to the molecular form of life than anything else, inhabiting directly the solid medium, exactly as we found them by the zillions inside the rock at the beginning of this book.

The crystalline form of life is associated with the solid medium, with rocks. In order for the field of the rock or crystal to exist, it needs energy, which the field can take from the temperature of the rock. Yet if it is too cold, it can take energy directly from the chemical bounding forming the rock, or from both, since this is mostly the case. The entire rock formation is alive, with the field using its particles or entire lattices to generate the resources and energy necessary to keep it alive. What form of life does the field maintain in this manner? Any life and any intelligence already existing there or only projecting. The rock can hold its own unique entity, living there for a long time, only to die currently for someone to make three extra barrels of oil a day through fracking.

Life from other mediums can be only visiting, projecting there only temporarily, or replacing completely the original unique form of life of that medium. It can be simply you meditating in your back yard projecting into the beautiful mountain, or it can be any of your old concepts from your childhood still there, decades later, still roaming around. There can be the dead soldiers from the last battle taking place there, almost two thousand years ago, still fighting, with the field outperforming itself only to make everything possible ever since, and with none of the soldiers ever suspecting being dead. There might even be entire long gone civilizations in the mountain and in the entire tectonic crust below, with their people still hanging around as watchers. Watching you explicitly right now, just as you watch your soap opera, while laughing and crying alongside you and your loved ones.

If you find it incredible how intelligent life is trapped in various mediums and lives there the same sequence of events endlessly, many times unknowingly, it is not exactly the original life trapped there, but only its projection, yet still intelligent. This is also your case. Your entire existence here in this world is nothing else but a projection, according to many schools of thought and religions. You might be trapped here similarly, living the same set of events endlessly. You can even be in a coma, with all these happening only in your imagination, so how could you ever know? Whatever the case is, the field allows everything to happen, and it goes through a tremendous effort to make everything possible, even if it has to take the mountain apart molecule by molecule while reshaping its entire bounding structure just to scavenge the necessary energy to sustain your life.

Nanobs are crystalline formations, byproducts of an intelligent restructure of a solid medium using the generated energy and supplies to reshape the field in order to hold realities in its matrix. Nanobs, through their structure and form, are the most efficient crystalline formation. This is why they were formed, since it took the least amount of energy. Nanobs are not fossilized, long gone living beings, but byproducts of living beings. However, if these nanobs participate in forming other nanobs, then yes, they are living beings.

Scientists still search for a presumed nucleus somewhere in the middle of a nanob, which is never there, probably because nanobs are not cells, they are not cellular or organic, and they do not even inhabit liquid mediums. You cannot even form cells at the nanometer scale, because molecules measure in nanometers, and you cannot make cells with only a few molecules.

One scientist was so desperate to find a nucleus and therefore to prove that nanobs are fossilized nanobacteria, that he distorted his recording to glow in the middle, and made it look as there is a nucleus in each nanob. I found one conclusive result, when a scientist studying nanobs touched accidentally a petri dish, allowing contaminating nanobs to change the structure of the dish itself, embedding her fingerprint in it. Because nanobs live in solids, and they were alive right then on her own fingers. How many times did you touch glass, to leave your fingerprints embedded in glass itself ever after?

The name nanob comes from nanobacteria, which is impossible to exist, because bacteria are of the cellular form of life and therefore must measure in micrometers, not in nanometers, because there is not enough room to hold the billions of cellular components making it possible if the cellular components themselves measure in nanometers. This is why bacteria are called microbes, because they measure in micrometers, not in nanometers. However, nanobs can be of the molecular form of life, since this measures in nanometers, yet it seems that whatever they call nanobs are the byproducts of the molecular form of life living in rock, while they ignore the living molecules themselves, the actual molecular form of life.

Nanobs are a unique form of life, crystalline life, found everywhere right now, even on Mars, carriers and maintainers of an ever-present life at molecular level, and could be of the molecular form of life. Nanobs cause matter itself to diffuse into life at very small scales, measuring nanometers. Let us now search more types and forms of life.

Nuclear Life

Life exists at all classes of organization, starting with atoms and molecules, and extending all the way up to the largest scales possible, with Life spanning the wider world to form, create, or give birth to the One. Let us descend now as low as subnuclear levels, to find more life.

Our only means of searching for subnuclear forms of life are to study possible traces left by life at very small levels. First, there is no reason to consider that the smallest form of life is ionic, since that might be only the formation and manifestation of even lower forms of life. We consider ions alive, yet everything is alive at the ionic level, forming the entire ionic form of life. This stands at the base of the molecular form al life, cellular form of life, organic form of life, and then social class of life. Let us look for traces of life now at an even smaller scale, right inside protons.

Protons are not as exactly as they are described by the current atomic physics, nuclear physics, and quantum mechanics. Protons are formed throughout space everywhere, mostly between stars, between clusters of stars, between galaxies, and between clusters of galaxies, everywhere, in extraordinary clouds of hydrogen, or protons. These clouds are formed slowly, throughout zillions of years, out of electromagnetic radiation, and when they reach a certain density, they collapse in themselves, they condense slowly at first and then very intensely, to form all the stars, with the Sun included. They form clusters of stars if it is a larger cloud, they form an entire galaxy if it is an intergalactic cloud, or they form an entire new cluster of galaxies if it is a very large cloud formed between clusters of galaxies. It is the electromagnetic radiation coming from all stars and galaxies forming these gigantic clouds of protons gradually, throughout the eons, one proton at a time, while life in many other forms of life and from all mediums creates or gives birth to all. Protons are ionized by the same electromagnetic radiation, and can interact with the field in this manner. It takes energy to form one proton, energy supplied in theory by all stars in the universe. Overall, the constant formation of protons matches the constant transformation of mass into electromagnetic radiation taking place through nuclear reactions in all stars and quasars.

This is a stable, equilibrated model of the universe, far more consistent than the officially accepted big bang theory. There is only one simple detail that this stable cosmological model will never clarify alone using physics, the conservation of entropy. Entropy, or disorder, tends to increase in the universe, through each event and occurrence, yet it does not, because Life decreases entropy in the universe, keeping it constant. Consequently, in order for this entire universe to exist, in order for protons to be created out of electromagnetic radiation coming from all stars and quasars, radiation must be formed through the annihilation of other protons, while life manages this entire living process.

There is life in the ever-present field formed by the ever-present mass and radiation in every way, while Life makes everything possible. This means that Life exists at all levels, at all possible scales, including subnuclear scales, and even lower. Furthermore, forms of life living at these very small scales make protons possible, and even manage the formation of protons in the entire universe. It is also possible that protons are not actually formed out of pure radiation, since protons are never annihilated during nuclear reactions, but they are only made to exist in a lower state spacetime continuum, through a process that releases a great amount of their energy, nuclear energy, the same energy transferred to the electromagnetic radiation to form, later on the new protons. In this manner, protons might not be formed straight out of empty space and new radiation, but they might result from a process of change of state of the spacetime continuum, process made possible by the field itself and by life. Because only Life is capable to hold entropy constant in the universe, and therefore only subnuclear life is capable to generate protons directly, to grow them from one state of the spacetime to another, or even to give birth to them altogether.

Living States of Water

We learn in school that the chemical formula for water is H_2O. This is not exactly true, but only statistically correct. The structure of water is different than H_2O, sometimes being HO, H_2O_2, or H_3O, but only locally and only occasionally. There are dozens of distinct states of water, not only ice, liquid, and vapor, with each one existing at its own distinct energy level.

Life is an intelligent scavenger of energy, and it certainly profits of this difference of energy coming from the continuous fluctuation of these distinct states of water. It is not much, which is not comparable to what it is generated by the exothermic chemical reactions allowing organic life, but it is energy, a different form of energy, and it can be scavenged intelligently by life and used to hold and maintain living intelligences in the field.

Who or what can use this energy? There can be unique forms of life living in water, larger or smaller individuals or entities, shapeless or not, perceived by us through eventual turbulences, currents, or differences in temperatures that they can induce. Mythology mentions such living entities. The field of water can also hold any projecting form of life. Anything and anyone can project in the medium of water, you, your thoughts, your grandma, your emotions, your concepts, anything. You cannot notice physically the difference in the states of water, however, you can always distinguish life in the universe from its characteristic and ability to increase order or decrease entropy.

As an example, large groups of people can control weather, creating rain and sunny days at wish, by controlling the states of water. Do the statistics, to find that there in an increased chance of having a nice sunny weather during weekends, vacations, and legal holydays, all thanks to children, since they are out of school and they are free to play outside. You can also notice the drastically increased order in the solidification of liquid water in the presence of abundant, positive thoughts and words. This can form very beautiful snowflakes during the first snow of the year, when all children run outside to play, while the snow is very beautiful. Yet this used to be the case decades ago, because currently, drugs, medication, and food additives stop children from projecting, and cannot form beautiful snow in the air, beautiful ice on windows, and beautiful snowflakes.

Currently, you can still find frozen water condensation on your windows forming beautiful flowers sometimes, and other times forming nothing at all. Just study it closely. Are you happy or sad then? Are you stressed or in complete control? Do you have positive or negative thoughts? Because everything is reflected and projected in all the mediums around, including in liquid water, snowflakes, food, drinks, and ice, while this is mind over matter at a very smaller scale.

You might have brought your own direct contribution to the above events, and never even knew it. Whatever it happened, it was always life making it possible, a different form of life.

Aerial Life

Just as water, air can hold life in various manners, according to its energy, or according to its local difference in energy, anything that can be scavenged by life and by the field holding its intelligence. However, different than water, air is considerably less dense, and therefore the field that it can hold is very weak, at a significantly lower resolution, rendering it less capable to hold a more diverse reality.

Currently, you still have your physical body to return to from your projections, while when you die, you are on your own. What could you do then? At your choice, you can either go ahead and move on with your wider existence, or you wait for your companions to finish their lives here and then move on together. You can return to Earth for another existence, you can join some nearby collective consciousness endlessly, you can find another giant reality similar to Earth, or you just find something else to do up there in your higher world where you live.

Can you go serve the Divine when you die? Yes, certainly, since you always serve the Divine, throughout your entire wider existence, here, up there, and everywhere else. You are always part of Life and the Divine, since Life and the Divine brought you here in this particular world as an intelligent human being, and therefore you must always serve them at your third intelligent level. Because you do not serve the Divine at your zero and first developmental levels, since at your zero level you have to follow addictions and entertainment, while at your first developmental level you serve the multitude of tyrants, dictators, profiteers, masters, ideologies, hierarchies, and the Consensual Matrix, instead of the Divine. This is how you fail Life and the Divine. You still fail Life and the Divine while living your life at the second animal level, because your meaning in life is not of animal nature, otherwise Life would have brought a fish or a horse in your place, yet they brought you here as an intelligent living human being. You truly serve Life and the Divine while living your life at your third developmental level, which is the intelligent human level, while fulfilling your intelligent needs and meanings at the human level as you do now while reading this book, and as you do while living your wider existence within normal, natural worlds and environments, and not consensually in the Consensual Matrix. The Consensual Matrix promises everything only for you to serve it, and this is how most of the wider world serves it eagerly, meaninglessly, and unfulfilling.

In the case of the aerial type of life, you can follow the energy in order to find life in the air, everywhere in the atmosphere of earth, scavenging weather-related energy, and creating all weather-related events as side effects.

Crop circles are events directly related to aerial life. Winds are relatively chaotic, following closely differences of pressure and temperature created by the difference of temperature in which the Sun heats the surface of Earth. However, when you project into rock, into the crust below, the living concepts that you create there are capable and very determined to hold consistency down to the slightest detail, since all life does so. Later on, when the same living concepts project into plants and air, the same consistency is projected, intact. All local weather elements follow and hold the same consistency in details, reflected in everything, including rain and wind patterns. This is how the image that you held strongly in your mind right before you went to bed materializes now on snow, on grass, and on water, or it materializes on some crop field hours later and kilometers away. It is also possible that while projecting into several stronger mediums, in order to enhance your mind's strength, clarity, and capacity, you can undergo a more conscious action of depicting what you imagine into shapes, materializing it directly into your new medium. This is more likely to happen if faint plasmoid spheres of light are present on site, since they are related directly to your projections. However, sometimes, these spheres of light are directly related to unique, independent aerial forms of life, while other times, they are related to projections coming from other rational species living underground, from lower rational species as cats and horses, or from alien species projecting occasionally into our realities. Why the similarity? Because it is the same life, regardless if it belongs to that medium or it is only a projection.

Currently, it is considered a psychic ability to act objectively in a subjective manner, which is mind over matter, to influence objective events with the power of your mind, or to bend reality. Yet there is more to state here, and I study it throughout other books of this series.

Orbs

Plasmoids are the form of life inhabiting the plasma of stars, quasars, fires, ionospheres, and even air and empty space, any cloud of ionized particles. There are various forms of life inhabiting plasmas, scavenging energy in various manners. Plasmoids scavenge only the Brownian motion of ions, directly related to the heat of the medium.

Plasmoids are spherical, luminous, colorful, and sometimes transparent, depending of where they live, while we call them orbs. Many orbs are associated with UFOs and extraterrestrials, which is not exactly the case, since we share the planet with them. Orbs inhabiting the ionosphere are very large, relatively short-lived, and are so large sometimes, that they can even be seen from satellites, if satellites exist. While the orbs from the air here at the surface of Earth are very small, made of white ionized fog, they manifest for a short time, and can be spotted in your regular pictures. Some of these air orbs are either unique individuals, or they are only their projections, which is the case with all orbs. When psychics act directly, consciously on the field, tend to form orbs and have them around, wherever they are.

How do you distinguish between unique individual orbs and projected concepts? Unique individual orbs tend to live longer, they are relatively less transparent, and they display more details. As an example, when people around start seeing a UFO, and if it looks as a plasmoid or orb, it can be the act of their own projection and not a real UFO. Soon, there will be more orbs forming not too far away, in a specific formation, moving together, in a larger group. What happens is that more people see the UFO, yet from different angles, and this is how they form identical orbs through projection, joining the others. This does not depend only on people's ability to project into the air or ionosphere and form orbs, but it depends on the medium itself, since mediums are more or less malleable, more or less full of life, depending on many circumstances. Mediums make you a better or worse psychic, depending on how alive they are, depending on how much life they hold, and depending on how intelligent they are. Strong electromagnetic radiation as wireless signals can destroy the life in a medium, limiting your ability to project and bend the field.

Nuclear explosions killed plasmoids in the past, and now it seems that orbs make sure that there are no more nuclear bombs detonated, by disabling them themselves. Psychic people can do so directly, while their higher ability is seen as an orb, which is mostly the case. However, orbs and the plasma field itself are attacked and killed systematically when they become too strong, too alive, and therefore too capable to enhance people's abilities. This cannot be done directly, since there are higher laws protecting all forms of life, yet the human mind is very creative, rendering everything possible, both good and bad.

There was a war in the Balkans decades ago, when they used dirty bombs to spread radiation all over the area, killing all plasmatic life in the process. Coincidentally, throughout time, this form of life had become so strong, that it interacted directly with the people, enhancing their awareness and abilities, in a benefic manner. Now everything is dead, with everybody back in line, serving society consensually and obediently. While now it happens in Ukraine, soon to come to a neighborhood near you.

Rockoids

Rockoids live life in the hard rock of mountains and of the entire crust of Earth as a variety of forms of life, with some already studied throughout this book. How can you tell if a mountain is alive and how alive it is? Solid life tends to overcharge and overact on the crystal lattice of rocks, increasing the electric and magnetic fields in the area. This field can be measured directly on the rock, and it can be felt directly, giving you the feeling of a wider mind or consciousness, of a wider certainty, and of a deeper perception.

It happens that living mountains can make your dreams come true, indirectly, since through your expanded abilities, you can bend reality. People do so, and many times, they do not even know it, so they interpret it in any manner. There are many living mountains, not only Mount Shasta, but Mount Dulce, along with the very old mountains of Area 51, all being able to enhance your awareness and abilities. Measure the field intensity in and around many famous sites as Stonehenge and the Pyramids, to find it intense, coming from the rock itself, from everywhere. If the living rocks were not previously preset there, then they were carried deliberately there, for long distances, for their living cognitive characteristics.

Why would people engage in such an irrelevant activity of hauling heavy rocks across the country for years or decades in a row? Everything relates with astral projection and even with hibernation. Because rockoids, or living rock, will allow you to project in them, through them, and through their entire planetary network of life. In this manner, they can open your way towards the entire wider world, everywhere you want to project, visit, and inhabit.

Very long time ago, people did not live in houses, even before Antiquity, because long ago, people were not forced to marry only one spouse, but they could live life in larger families spanning entire areas, as these become intelligent human environments. When your environment is natural, alive, and of an intelligent and psychic level, everything allows you to develop to the intelligent human level and further, to the intelligent psychic human level, allowing you to develop your higher abilities, one of them being conscious astral projection. Larger groups of people do not live in houses and little bedrooms where they sleep alone, but overall intelligent human living allows you to be always with your loved ones, day and night, in very large numbers. What do you always do with all your loved ones, mostly when you are developed at the intelligent psychic level, and can make use of very high cognitive abilities? You can do everything in the world as everyone already does, but you also daydream together, while forming entire, beautiful, complex worlds, exactly as this world, or even more beautiful and more harmonious, everything that you ever desire. This world is made in this manner from the higher world by billions of souls. You can project together everywhere you want in the wider world, to be in any other world you please, or to be anyone and with anyone you please.

What exactly do you do in the wider world while you project there along with all your loved ones? You certainly learn and experience everything that you want, but mostly, you want to daydream there alongside billions of higher living beings in order to live entire lives with them, throughout higher worlds, life after life, however you want, and as anyone you want.

You can achieve these not as a soul, but as a living human being, right now, here in this world, but only if you are intelligent psychic, if you have the necessary mediums to help you project, if you have all your intelligent psychic loved ones in very large numbers at your side, and if you know how to project. Yet with an entire Consensual Matrix watching you closely never to achieve all these in any manner, it never happens, so you drink and smoke some instead, as everybody else.

People actually lived life as intelligent psychic living beings in very large families spanning entire regions of Earth, thousands of years ago, before Antiquity, or even during Antiquity. Because ever since the Consensual Matrix and the hierarchic Brotherhood took over Mesopotamia and the entire world, now you have only rigid laws, everlasting delusional ideologies, mostly social, political, and religious, and continuous discrimination and tyranny, either covert or in the open. Is it worth it? Because ever since Antiquity, you can never have actual human development in all domains, including art and knowledge, without higher abilities, without higher interconnectivity, and without astral projection. Is it worth it?

People did not live in houses long ago, before Antiquity, not because people had not invented the house yet, but because larger, overall families could not fit in houses, and because people could not heat the houses during the long winters of the ice age in Central and Western Europe. People lived underground and in caves in the mountains. How do you dig larger, sustainable habitable caves underground, without the ceiling collapsing eventually? You need dolmens, which are stronger pillars, meant to hold the caves stable. These are massive poles made of rock. You still find these everywhere, the well-known circular formation of rocks that stick up the ground, with no one knowing what they are. These circular formations of dolmens were once underground, holding the integrity of very large caves, inhabited by large groups of people, as they lived their life normally, naturally, and not consensually.

If you live there for millennia, in perfect, natural interconnectivity, those rocks fill up with life and intelligence, from you and from all your ancestors, within specific inner realities that are your home, from generation to generation. What do you do when you have to move, for various reasons? You take your holding rocks with you, the dolmens, since they have your worlds and realities in them. You have to carry them even for hundreds of kilometers, regardless if there is similar rock everywhere, because that is your true home, and there are your ancestors, along with your living culture and all your achievements, in those other realities. It helps you astral project while you hibernate, and you get to do everything you want alongside your loved ones even during the long winters, and not only during summer.

Because this is your life, now truncated entirely, as the Consensual Matrix had rendered it, while it made all the difference in the world. Currently, scientists find these circular holding structures used in holding habitable caves, they find them at the surface as in Stonehenge, Malta, Middle East, West Europe, and Turkey, or they find them deeply buried underground as in Turkey, and they can never figure out what they are. Because when scientists see these rock formations out in the open, they can never figure out that only millennia ago, these structures used to be the dolmens and the arches holding wider underground caves. They even find them buried underground in this exact form and alignment, and they still cannot figure out what they are, but they only claim that primitive people buried them there, for a later retrieval, to use them for specific ceremonies and other superstitions, as these take place throughout the hierarchic Brotherhood. Study closely all pyramids in the world to find large networks of caves and tunnels underground, with all the rocks in the area showing a strong electric and magnetic field, while many times, all cave formations date further in the past than the pyramids themselves.

Why are rock formations more or less full of strong, living electromagnetic field? It always depends on their position, on their age, and on the people living there in successive generations. When you love something or someone, you transfer your feelings, field, projections, energy, and life into that someone or something, both consciously and unconsciously. When beings, or large groups of beings transfer living, benefic concepts into any medium, as mountains, air, Sun, planets, or other living beings, they enhance the life, field, and abilities of that specific medium. Happiness, harmony, and love can make everything possible, through the accumulation of love, life, and more. These specific mountains could have been charged in this manner for millions of years, while currently, most of them are restricted areas, holding military bases and underground cities for the Upper Brotherhood and the Elite, on top of a honeycomb of other cities dating back eons, formed by forgotten civilizations of Earth, with some of them still inhabited.

Why living there? To cherish the place some more and to give it more love and life as long as you are alive, or to draw out that life altogether and to use it in a very egoistic manner. Everybody wants life, unlimited quantities of life, along with the possibility to exploit the field directly, while bending reality, and while turning the entire human timeline in your favor, while this is what the Consensual Matrix does here on Earth. With the entire Brotherhood working overtime to make it possible, in the detriment of the Masses and the Brotherhood alike, while this is humanity altogether. Humans against humans, rocks against rocks, and therefore life against life. Yet you make twenty more dollars in profit this month if you only serve even harder and if you exploit even more. So yes, how lucky you are to be alive currently.

The One

The One is alive, and you can check all our definitions of life to verify it. The One is a unique form of life, and will always remain unique. As seen, we take the One apart into its Life, Consciousness, Interconnectivity, and wider world, only to understand it, since it is infinitely vast and complex.

You can raise your perspectives if you can, all the way to its level of existence, in order to catch a glimpse of how it is to be the One. At that higher perspective, you notice that you are everything, with everything being part of you. You notice that there is no outside anymore, since the concept of outside has no meaning now, everything being inside, within yourself, while you are also present everywhere. It is the same with time, because at the comprehensive level of the One, you exist continuously at once, and it lasts endlessly. You are alive, and your life is the entire Life, everything ever alive, everywhere, in all forms, and in all realities. Your intelligence is the overall Universal Mind, held by the universal field, present in everything alive. Through it, you can do anything that you ever intend, since you are everything. You are omnipresent, omniscient, and omnipotent, you are everything interconnected, you are the One.

Without a constant passing of time but with all time passing at once and lasting endlessly, the causality is different than what we experience at our regular scale of the world. We experience events here as sequences of cause – effect – cause – effect, while at the highest perspective ever, causality tends to lose sense or it even reverses.

Is the One the Deity? It might be, since there is nothing above it, it is as far as you can perceive and understand.

6 THE ORIGINS OF LIFE

There is one way for life to have appeared on Earth. You know it well, and you accept everything according to the truth and to your ideology. Yet if your mind is free of ideologies, and if you are eager to learn and understand more about life, including its origin, then you are in luck. If not, if you prefer to hold on to your ideologies including science, politics, cults, and religion, then you can do so. I will present here not only one origin of life, but seven. However, only one is true, or none at all. While you cannot take sides, you cannot debate, but you can use only rational thinking, at the third intelligent human level, to understand life starting from its origins. How did everything start? Spontaneously, or not.

Life appeared spontaneously, out of ordinary dirt, and this is called Spontaneous Creation. Yet if you were alive over a century ago, not only that you rode a horse to go to serve, but this is what you learned and believed in with all your heart: spontaneous creation. Was it the intention of the Deity to form life spontaneously, or was it simple probability at work? Everything depended on your culture, religion, background knowledge, awareness, level of understanding, main intentions, ideology, family, and even economic interests. This is still the case currently, and everything adds up to form not only the current biology, but the entire current science. Scientists of that time went as far as proving that worms, flies, and even mice came to life spontaneously overnight, out of simple dirt and water, and everybody believed it. It might seem incredible currently to believe that people were so ignorant over a century ago, to assume that mice can appear overnight out of moist dirt, but they did. Did they really? Just consider what people believe in currently. Big bang, random evolution, life exclusive to Earth, and humans the smartest, the best in the universe. What is the difference between these and the spontaneous creation? There is no difference, since these are beliefs, first level consensual thinking, and therefore everything is the same.

We have seen how, for centuries, science attacked religion in every manner, throughout all its discoveries, concepts, and theories, many times inaccurate, fabricated, but maintained valid consensually by a global consensus including all educators and scientists of the world. Science had an agenda then, the same agenda that it has currently, and it went through everything possible only to implement it. This is why you cannot consider isolated cases to understand science and religion, as the spontaneous appearance of mice or the first man on the Moon, but you must study closely everything that all the powerful social factions tried to do and achieve through all consensual scientific knowledge that they maintained instated, including creationism, spontaneous creation, and the big bang theory. This is the case not only in science, but in politics, business, education, and media, since social competition takes place globally, which is deeply embedded in everyone's needs. Social competition was the characteristic making our robots learn, survive, adapt to the environment and mostly to themselves, become alive, and then even morph into organic life. However, only underdeveloped people engage in social competition, while when you study animals, you find them interacting harmoniously even in the jungle.

Was the world created by the Deity? Was life created by the Deity? We have suspected many times that we live in a simulacrum, a reality created only for us to exist in and enjoy.

This is not unprecedented, since we find many created realities everywhere in the wider world, with some being created by humans, and even by you, as all your mind realities, throughout your learning, reasoning, and daydreaming. Therefore, this world should be similar to these. Furthermore, our creators could live life in a created reality themselves just as well, reality made only for them to exist and enjoy.

It is repeatedly stated throughout old records that humans are made in the image of their creator, while Earth is made in the image of the higher world. As above, so below. Here are some possibilities: life was created exactly the way it is currently, life was created and allowed to developed, or life was not created, but it simply entered the newly created reality from the upper reality of our creators.

According to scriptures, all the above are true, simultaneously. When you live in a created reality, not everything that you learn and perceive there is true, because created realities try to mimic the upper reality, only to make them more credible, more adequate for learning and reasoning, and more enjoyable. Furthermore, all created realities have their limitations, for all reasons, as cost, efficiency, time, space, codes, and laws. This is why every created reality that you are familiar with, including the "World of Warcraft," the worlds of "Need For Speed," or the holodecks of Star Trek, exist in a limited space and time, among others. In this manner, only what exists objectively in this world, everything tangible can be real, with the rest, everything further away being an illusion.

Technically, you can never know how far this world lasts. You are from India, and you learn of Africa and America in school, but do Africa and America really exist, or they are just an illusion, a concept, meant to make this world more credible? If you have already been there, then yes, they exist. Yet how far does this world last? How certain can you be that the Middle Ages really took place? In Star Trek, the holodeck changed its point of origin by creating and recreating its environment with you always in the middle, always keeping its boundaries only several feet away, with the rest simply projected on walls as in a movie. We can see the Sun and the Moon on the sky, but do they really exist, or they are simple projections on a dome? If you play "Delta Force," or "GTA," you know exactly how these limitations work, and how you are always constrained to remain within your videogame world, in every way. How can you be sure that anything from beyond Earth is real? Have you ever been there? Is it true that they exist only because science states so? Because religions state clearly that the Deity had created the Earth and placed it right in the middle of the firmament, fixed, with the stars, the Sun, and the Moon only to rotate around and give us heat and light. Does it seem familiar? Could it be true? No, not according to the current science. How much of the current science depicts the truth, and how much is only consensual? Study religion closely, to see how everything depicts a limited, artificially created reality, made to fulfill a definite purpose. Study the very old spirituality closely, to find it stating the same.

Does this mean that everything that we learn in school is irrelevant and we should turn to veneration right away? Do as you wish, yet veneration is first level consensual behavior, but not intelligent human meaning and fulfillment. However, everything manifests in our world according to our Creator or Creators, according to their plan, intentions, and reality, and according to the wider existence in general, according to Life, and according to us, making us simply co-creators in a smaller or larger manner, throughout this extraordinary existence. This is the case everywhere, in all realities, and not only here, going all the way up to the Ultimate Creator, the Deity. Yet there might be only one created reality in the wider world, ours, in which case, the wider world is simpler than what we envision. Furthermore, it is possible that this world is the wider world itself, more or less limited, while science had stated the truth continuously. Note that the origin of life is in complete concordance with the nature and characteristics of our world, following it closely.

The divine creation of life has been challenged by an old encounter of ours, the theory of evolution. Which one is true?

We cannot know, because we were not present when Earth and life were created or when they appeared spontaneously or not, and this is why we have to learn everything from ideologies, while these offer only first level knowledge based on beliefs. This is consensual knowledge, not necessarily accurate, despite of what all ideologies might claim, while these count in tens of thousands, many times fighting and contradicting themselves. Ideologies function in this manner through beliefs, and so do religions, along with the current science itself. The current science is a scientific ideology, based on a scientific consensus. Everything voted, consensual, and decided for one reason or another is a belief, and not necessarily accurate knowledge. While all accurate knowledge is unique, correlated with the natural laws of the universe.

Science could have never challenged religion throughout the past centuries, mostly through its main strengths, the theory of evolution and the big bang theory, if it was not an ideology itself. We have studied these two theories throughout other books of this series "Human," and found them irrelevant, along with most of the current science.

Therefore, we might never know the true origin of life on Earth according to what the Masses know and according to what the Brotherhood are told, for various reasons, but we can always know it according to what other civilizations know. Because even though it is always a belief, most of them might have been there, or here, when it happened.

The question is who to believe. If you are a freethinker, free to analyze every theory, supposition, or belief, then you are in luck, because there are many people, mostly in the upper society, forced to abide to very rigorous sets of consensual knowledge and beliefs, according to their faction, and they are not allowed to think and learn anything else. This is also true on the bottom society, not through extrinsic constraints, but surprisingly, through intrinsic means. Even when you are genuinely free, you can always succeed in imprisoning yourself through your own ideology and strong personal convictions. This is why we are willing to accept one alternative of origin of life over another, not necessarily because our background knowledge, perspectives, or means of understanding are limited.

Intelligent creationism is another theory of the origins of life, highly debated in the current society, not necessarily because it is more or less accurate, but because it is not conformed, and it does not apply to the Divine creationism, neither to the theory of evolution. What the intelligent creationism considers is that Life is too diverse and too intelligent to have been created through simple accidental evolution. There are distinct characteristics of life stated in the intelligent creationism theory, as the genuine electric motor rotating the flagellum of a bacterium, technology that could have never appeared randomly, since we use the same technology in racecars, and we have invented it intelligently. This is an error of reasoning.

Why is the Deity incapable of such a technology? Probably because religion opposes science currently, and therefore the Deity should also oppose science, technology, and intelligence. This seems amusing, but what else could the intelligent creationism theory state in order to detach itself from the Divine Creationism?

Why do theories of origin of life have to detach themselves from other theories of origin of life? By definition, 'theory' means speculation, assumption, or possibility, but not accurate truth, which means that theories are not accurate, but only consensually valid. Theories always struggle for distinction, uniqueness, popularity, and credibility, but not necessarily for accuracy. Furthermore, mainstream theories have other meanings than informing the people, since they have to indoctrinate the people in specific manners, exactly as ordered. It takes several alternative theories to indoctrinate the people, only for the people to have a variety of indoctrination to choose, because people still tend to be different, and therefore they have to be reached differently, which makes the job of all social engineers more tedious.

Which theory of the origin of life is true? You choose, yet

we have already encountered this topic. Life is everywhere, in all types of life and forms of life, and in all worlds and realities, while you cannot evolve from the nonliving to the living. Which means that there is no origin of life in the entire wider world, because Life is always alive in the wider world. There is no origin of life, since life is eternal in our wider world by trivial default, since Life is the wider world itself. The wider world itself is the objective, existential perspective of Life, while Life is its living, empirical perspective. You cannot have an origin of life in the world, since life is the world, while the world is alive.

You cannot simply choose your theory of the origin of life from among all the theories in the world, as you choose your favorite sports team, because accurate knowledge cannot be chosen since it is already unique, while in this case, it relates directly with life and with this world. In contrast, all first level knowledge is consensual, judicial, and ideological, and you only take a stand in an unending debate, while debates themselves are first level common thinking. You cannot reason accurately through consensual knowledge or beliefs, since they are not accurate but deceiving, and because no one listen to you anyway.

This is the case with every ideology, and this is how they get you. The intelligent creationism theory states that life had appeared on Earth in an intelligent manner, designed in this manner by an intelligent being. This being can be any intelligent individual, from aliens to smart humans and higher beings. We have seen throughout this book that the organic form of life is only one form of life among many, coincidentally the most adequate form of life matching this environment, yet it is only a specific mode of Life. Coincidentally, the intelligent designer from this specific alternative theory of origins of life has designed life to exist and behave in this exact most efficient way in this environment. Furthermore, it made it adapt and reshape itself to cope with every change in the environment of Earth, doing everything down to the slightest detail.

Let us study a fourth alternative of origin of life on Earth, Panspermia. Panspermia theory states clearly that there is no need for a creator, intelligent or divine, since Life always seeds itself from one stellar system to another, form one planet to another, in every possible manner, including through interstellar travel. There is objective life in space capable to reach any place of the universe by floating around or by taking a ride on any space object, intelligent or not. All spores can stay alive in any environment, including space, until all conditions are adequate to sustain life, and that is when they hatch. Comets are formed mostly of water, and they contain frozen life of all type, including germs. There is a perfect correlation throughout time between major plagues and the apparition of comets. Panspermia allows life to be spread, accidentally or not, only by intelligent life, making Life possible in the universe in every manner.

This seems accurate, and it could have happened here on Earth. Yet how exactly did life originate in the entire wider world. Probably accidentally, exactly as the big bang itself, since the panspermia theory cannot answer this main question. Which means that, if the panspermia theory of the origins of life cannot actually explain the origins of life, it is not exactly a theory that explains the origins of life but something else, something related to the transportation of life, and we should not waste our time.

Yet this theory still works in a created universe, with life created simultaneously with the entire universe, or with life found everywhere in the universe, which contradicts the idea of an apparition of life, if life is already there everywhere. No theory of origin of life states that life is already everywhere, in all forms and realities, but they struggle to come up with a speculation of how life appears or is created everywhere, relatively recently, to match science and the entire Consensual Matrix.

Could Life have appeared accidentally in the universe? Certainly, since there is a probability for everything when you can never know it for sure, since you are never as old and as large as the universe, and you cannot witness it accurately firsthand. This is why you have to accept this rigid pathway of causality in the universe: space creates matter accidentally through big bang, then matter creates life accidentally through the theory of evolution, and then life creates intelligence accidentally through the same theory of evolution. While we found intelligence to be the essence of life, or to be life itself, but not its product. With the field to be the holder and maintainer of life, intelligence, and all realities of the wider world, making them equivalent.

At a higher level, this rigid causality promoted by science is called 'Order out of Chaos.' Implicitly, Life had evolved from simple matter, chaotically, as a tornado going through a junkyard forming chaotically a brand-new jumbo jet. This is possible, in an infinite universe, since life has an infinite time to appear in this manner. Who could validate this alternative anyway? Intelligence could, certainly, since theoretically, it precedes life, yet we have to involve the field in our study. Can we? No, or at least not according to science. Science studies the field in order to produce light bulbs and electric motors, which is good, since humanity had spent too much time using candles and horses throughout dark ages, when mainstream ideologies banned reasoning and therefore technology.

The field is everywhere, while life can exist entirely within the field, without even employing field tools or living beings to maintain itself within the field, to shape and reshape the field in an intelligent manner. While as you study closely all realities of the wider world, you notice how they all have a cognitive nature, similar to all mind realities that you must form right now in order to reason intelligently alongside this book, while they all have a creator, similar to you while creating all your inner mind realities, while daydreaming and while reasoning intelligently.

Furthermore, you never create the reality first, then you furnish it with all the necessary objects, and then you bring in life, but you create all these along with the entire reality, which is the case in the entire wider world. You never have realities appearing first, followed by all matter within, followed by living matter consisting all life, and then followed by unconscious cognition and then by intelligent one, but they are all part of the entire world continuously. You can imagine anything you want, including entire deserted worlds voided of life, yet you must always maintain correspondence with the real world, otherwise you lose touch with the real world, while causing all your inner mind worlds to remain unsuccessful throughout the fulfilment of all your needs and meanings.

Why exactly having life in the universe? Why not having life only within the field? Because living beings are capable to act on and reshape the field consistently stronger and more precisely, allowing higher forms of life. Higher forms of life are living beings, and this seems to be a main purpose in life, development at all levels of life.

The field is everywhere, which is therefore capable to form and give birth to life everywhere, in every vacant environment, world, and reality, through every vacant niche, as overpopulated as it can be. The field or the intelligence held within the field, can start life from scratch, in the most efficient form. While the field can destroy any form of life when it is not capable to adapt and cope anymore with its environment, morphing it entirely into a completely different form of life, while switching Life from one type of life to another and from one form of life to another if necessary.

How exactly can the field do so? The field itself is alive, which is only a matter of rendering its life at the necessary form and developmental level, matching the environment, which is also alive and part of the field, sustained by the same field, through similar intelligences. In the next chapter, we see how the field creates organic life from scratch, and how it guides it throughout billions of years to the outstanding level of strength, intelligence, and diversity that we are, experience, and encounter currently.

Are these theories of the origin of life true? No, since they are only theories or assumptions, exactly as they state in the title. Study them at your own risk. The environment of Earth has changed radically and repeatedly since the Earth was formed. The intelligence of the field gave birth originally to a different form of life, Plasmoids, or ionic life, then to Rockoids, since liquid water was not possible on Earth to hold organic life. The field had morphed and developed forms of life repeatedly before morphing into organic life, and then taking it further to larger forms of life interconnecting entire biosystems, as Mother Earth.

The life and the essence of life in all its previous forms of life on Earth are still there, in the field, at the base of organic life and everywhere else. However, I will keep this model simple, considering that the intelligence of the field has created organic life from scratch, and we do so as a simple study.

How could the field ever achieve life altogether, including the entire electromagnetic type of life with the organic form of life included? The field is alive in itself, filled up with living intelligences everywhere, as raw as these can be in the field. Because within the field, you do not exactly need the rigid, material, objective bodies to hold intelligences, since all vertices of the field, smaller or larger, can hold through their own living, intelligently modulating anything, including entire worlds and realities filled up with all possible living intelligences, and this is Life. While everything else found on larger and larger scales and throughout higher and higher forms of life and classes of life are simple expansions of all these raw vertices and raw intelligences of the field, yet still alive, aware, conscious, and intelligent, with you as a living human being included.

The field does not really act alone, rigidly, following the laws of physics, or it does so from a strictly technical perspective. While the field uses intermediary procedures of perceiving, living, and functioning. These are realities, they are in all life and intelligence, and they sustain themselves and interconnect life and intelligence. It is similar to the software that you use in your computer to help you rewrite the machine code throughout all your tasks. In this manner, through all its realities, the field allows all life in the universe to participate in the creation or in the birth of new forms of life, throughout new environments. With the same raw intelligences of the field acting and reacting throughout all these higher worlds, forms, classes, and realities, and they do so by gathering by the zillions to form entire cognitive systems themselves from the bottom up, many times just as specialized but significantly more capable than the individual elemental intelligences of the field, now acting throughout all upper forms of life and classes of life.

How new exactly are the forms of life, environments, worlds, planets, and entire realities? They are new for specific places, yet they are not exactly new from a universal perspective. This means that organic life exists in the universe where the environment holds liquid water, and constantly participates in the creation and maintenance of new organic life from similar new environments, wherever the conditions match whatever we have here. Yet as always stated, you can never know how large realities actually are, since if they are only as large as those worlds and habitats, then there might be nothing past the lower orbit of Earth. Earth might not have to be entire and spherical, but only a larger chunk, wherever you happen to live.

Is the field now the creator of everything? Is the creator of everything intelligent? Intelligences always create their own inner realities for cognitive purposes, since this is how intelligences think and exist. As seen, you the conscious intelligence create your own conscious inner replica of the world where you learn and reason, and where you exist yourself. This is the case with all intelligences, and therefore with all living beings. Everything is possible through the field, to the point where you cannot distinguish anymore between life, field, intelligences, realities, consciousness, and interconnectivity, everything being one.

This is the answer to the origin of life, since life has always existed at the level of the wider world, since it is always part of this oneness, it is the wider world. However, locally, life will always exist on various forms of life, many manifesting at molecular, ionic, atomic, or nuclear level. Yet if it is not organic in nature, life might not be accepted as life by the mainstream authorities of Earth, and now this is what you have to believe.

When exactly did everything happen? Time is part of this spacetime continuum, while all continuums stand at the base of realities individually, and cannot be defined outside them. Therefore, terms as origin, age, time, and periods have no meaning outside our world. Similarly, you cannot distinguish between life, intelligence, field, realities, and interconnectivity, since the difference between these stands at a comprehensive level, and cannot be defined within each reality individually. However, since your life is multidimensional, transcending realities in large numbers, you can certainly answer this question yourself, because you are more meaningful and more capable than entire worlds and realities. This is how all the accurate knowledge is already in you. Yet to lower yourself to the erroneous knowledge filling up the mainstream of the current consensual society is a significant downgrade of your own nature and meaning in life and in the wider world, and you should be more careful and more responsible. Yet this is how living beings are exploited and even exterminated, through the lack of self-authority and self-responsibility, and it certainly works.

How exactly did the organic life appear or was created? Let us see.

7 MODEL FOR THE ORGANIC LIFE

Let us start our model for the organic life from scratch, in a bare, liquid environment, and just wait for life to appear out of randomness, as the spontaneous creation and the theory of evolution state. Because once organic life appears overnight, directly in form of flies and mice, the theory of evolution takes over and evolves the organic life further, one species after another as these fight for survival, accidentally. We already know it well and I do not have to write this entire chapter while you do not have to read it, since you have already memorized science in school word for word.

Or instead, we might consider that an intelligent designer, far superior to us, had built organic life from scratch, because it seems to include living structures resembling the old technology when this theory had been instated, as the old electric motor. In this case, we can always employ science to model organic life for us throughout this chapter, and rewrite biology altogether. This is what happens currently, with authors and scientists going with or against the scientific stream, while trying to explain life locally, technologically, and always algorithmically and therefore mechanically, in every way it makes sense empirically to them and to the world. Because in order to prove the fork, they make parallel cuts in a spoon, and this is the ingenious proof that forks appeared out of spoons, thousands, millions, or billions of years ago, whatever number is more impressive to the public. People watch it in the media and wonder in excitement while having fun alone at home, and this is life. How exactly this type of life has appeared on Earth? It might not be the case only on Earth but throughout the entire Consensual Matrix, since this is the ignorance that you get everywhere in the Consensual Matrix.

Let us create a model for the organic life. We already know that life spans the wider world, while consciousness and intelligence span life and the wider world. We already have models of the universe in all realities, models of intelligence, and now, along with this model of life, we can integrate these into a main model defining our wider existence, integrating it into a one, unique tenth level concept of everything for the One, the Supreme Living Being, the Universal Mind, Life, Interconnectivity, the Field, Existence, or Causality, since you can call it as you please. Yet it will always be at the tenth supreme level, so good luck to us at our third intelligent level or even at our fourth intelligent psychic level if this is your case.

It takes us dozens of books of this book series "Human" to reach Life, as we need this model of organic life to understand ourselves, since this is our own conscious, organic perspective to understand everything. Yet it does not really matter how much we strive for accuracy, because from our own designated third intelligent human level, we will always perceive the elephant one bodily part at a time, due to human limitations past the third intelligent human level. Which is not too fulfilling, as we want more. Yet even so, let us be thankful for everything that we manage to learn and discover here at our third intelligent human level, because below the third intelligent human level, life and existence are rather chaotic and even irrelevant, just look around to see it yourself.

From the six models of origin of life presented in the previous chapter, which one is true, which one is accurate enough to get us started? None. Religion and spirituality might be exceptions, yet these are based on believes, just as science and the rest of the mainstream, and therefore just as in all our models, we cannot employ beliefs, ideologies, and any consensual knowledge, through lack of compatibility with the third intelligent human level. However, we always hope to find results consistent with religious and spiritual beliefs, strengthening our judgment. The other five models of the origin of life are useless, since they apply to a dead universe, with life present either exclusively on Earth, with life present only in a few places in the universe, or only temporarily, wherever and whenever the environment resembles ours. While this is not accurate according to all our models from this entire book series "Human." Again, we are on our own in this model of organic life, since all these famous scientific theories place themselves in a little bubble as wide as Earth, with life lasting until the following larger cataclysm.

While in reality, the wider world is teaming with life, continuously and in all forms and realities. Life creates and gives birth to life continuously, everywhere, and nothing is ever created out of nothing, but it is always born from life, with life, and within life, always to make possible even more life, everywhere, as much as possible, even with life on top of life, as it is the case throughout all forms of life. Life can never appear in an infinite universe, but it is always present, as infinite as the universe. Life adapts, reshapes, and morphs itself to be what we see currently, and to continue in this manner long after we are gone.

As we have already seen with the chicken meat from the grocery store. Because all meat from the grocery store is still alive for some time, not in the organic form of life, but in the cellular form of life, soon to die once again while decaying to the molecular form of life if the worms and maggots do not eat it first, drying up to dust if left on shelves for many years and decades. Dying down one form of life under another, down to the raw field, but always alive. Yet this is the case only in theory, because in practice, there are always living beings eager to eat the meat, reintegrating it in the organic form of life.

Life eats life continuously, keeping life alive at the cellular form of life of all bacteria, or even at the organic form of life of all little worms, flies, and maggots eating it continuously, to the point where you cannot have environments void of life, as long as the niches of life are still present throughout the environments of the world.

Furthermore, you cannot even create dead worlds and dead realities within Life, since you always end up with living creations, because you are alive. You might be able to create artificial dead worlds by using technology within Life, as even these are always alive in an artificial manner, which is the basic artificial life.

This is why organic life cannot appear out of nothing on Earth, even accidentally, because only the organic form of life can give birth to the organic form of life. While if the organic form of life is not present anymore in the environment, then the cellular form of life can make the organic form of life possible throughout the normal development of life, and only if the necessary environmental conditions are present, yet it takes a long time.

Furthermore, if throughout very unfavorable environmental conditions, all organic life dies out within any environment, then it only dies down to the cellular form of life, leaving only bacteria behind, and then leaving only viruses behind further down at their own molecular form of life, down to the raw electromagnetic field. Because every time the niches of life become unavailable through any unfavorable circumstance, the electromagnetic type of life dies down one higher form of life down, even all the way to the raw electromagnetic field, which is always alive.

Because if the entire world or reality remains meaningless, unsuccessful, and therefore unfulfilling for Life, then it shuts down, leaving only the higher worlds in place to fulfill Life. If possible, and whenever and wherever it is needed, Life always makes possible newer and newer lower realities even in very large numbers, always specialized, and always meaningful.

Because you can never simply start with an environment

void of life, to make life possible there through any ignorant theory of the origin of life as we see everywhere throughout all these religious ideologies, spiritual ideologies, or scientific ideologies, but you have to grasp the entire accurate truth of life, exactly as it is in the wider world.

Earth has several forms of life in parallel with the organic form of life, forms of life that are still present on Earth currently: rockoids, plasmoids, orbs, living water states, living air states, ionic form of life, molecular form of life, and cellular form of life. The field itself is alive, intelligent, and certainly interconnected, to be what we have currently everywhere. The living field of Earth was already vigorous and healthy well before organic life, yet organic life made it even stronger, extremely strong, making Mother Earth a very active living entity within our stellar system and within our entire galactic sector, if these actually exist objectively as seen in the sky.

Organic life had appeared on Earth once the Earth was able to hold its medium, liquid water. Even then, organic life did not appear out of nothing, but it had morphed out of the previous forms of life, ionic, molecular, and cellular forms of life, always standing at its base, still around currently through the field. Life is always there, it cannot have an origin, while it morphs and switches casually from one form of life to another and from one type of life to another, depending on many outer and inner circumstances, as it always does throughout the field. While the field always switches throughout all its states, including its well-known electric and magnetic states, according to many circumstances.

Throughout this book series "Human," we find that the field is related to cellular and molecular intelligence, while it is also related to computer worlds and computer software. We could never find an example where life was constrained entirely in a single reality, since each living being is alive at the confluence of two or more realities. Furthermore, when we have created our model of life using robots, we could not constrain life to our world in order to study it better, but we had to consider inner realities necessary to hold our intelligence. For each reality, we found that life is the summation of matter and energy capable to sustain the system's intelligence and to maintain it alive.

Let us start the model of organic life from scratch, over four billion years ago just to parallel science, regardless if it is the actual age or not, when liquid water is said to have first appeared on Earth, for the first time in its existence, since Earth had never been cold enough before to hold liquid water. Liquid water has a lower energy state. It had been formed, which is still formed by Life, since Life is an intelligent scavenger of energy, and it certainly takes her share from the bounding of hydrogen with oxygen, as it is exothermic, administering it molecule by molecule.

What makes the liquid medium so valuable for Life? What can she scavenge here? All chemical reactions are possible in the liquid medium, not much in the gas medium, and very little between dry solids, because it takes a very long time for chemical reactions to occur in the solid medium, and even then, they occur only locally, on the boundary surface between solids, and in very low amounts. In the liquid medium, all chemical reactions can take place, discharging high amounts of energy, more than enough to keep the field functioning at levels high enough to maintain consciousness, and more. While this cognition is of a very high resolution, perfect for Life. Yet highly energized plasma can maintain intelligence just as well, at even higher levels and resolutions, yet liquids can do just the same, in between stars, maintaining and conducting the field from star to star throughout the universe.

The field exists even in empty space throughout its very weak plasma, maintained by the few hydrogen ions per cubic inch found everywhere between stars and between galaxies. This is the same ionic form of life present on Earth, also present in fires and in stars. Plasmoids are also of the ionic form of life.

In space, hydrogen ions oscillate in the overall electromagnetic radiation coming from all stars, giving them enough energy to act and react in the surrounding electromagnetic field in any lively intelligent manner. It is the same on Earth, because the Earth temperature at its surface can give all ions enough energy to oscillate in any living intelligent manner, while making possible the entire ionic form of life found at the base of the entire electromagnetic type of life on Earth. Yet the entire energy of Earth comes from the Sun, we can still state that the entire ionic form of life is fed by all energy coming from stars. We start in space our model of the electromagnetic type of life, yet it is similar on Earth everywhere, within all molecules and within all cells of all organisms. However, in space, with only one proton for every cubic centimeter, it is easier to understand life as it takes place right on top of the electromagnetic field.

Even if there is very cold in the empty space, all protons still oscillate, at a temperature of about two or three kelvin. This oscillatory motion is always accelerated, while an accelerated charge in the electromagnetic field produces electric and magnetic fields, acting and reacting in this manner continuously in the electromagnetic field. Without life around, all oscillatory motion remains random. Yet with life around, all oscillatory motion of any charge in the electromagnetic field is lively and intelligent, following a specific encoded modulation that makes all life possible.

There are two correspondent oscillatory motions that you must consider in the electromagnetic field. The first one is the physical oscillation of any charged ion in the electromagnetic field, and the second one is the continuous oscillatory shifting of the electromagnetic field itself from its electric state to the magnetic state and back with each physical oscillation of the charged particle, because each physical oscillation is an accelerated motion, capable to create electric and magnetic fields itself, while in this manner acting and reacting in the surrounding electromagnetic field. While if these two oscillations are encoded in a lively intelligent manner, you have life, because this encoded oscillation becomes an electromagnetic matrix capable to hold within an entire inner reality, capable to hold intelligences, while these intelligences coordinate the entire oscillatory motion from the outside world in a specific intelligent manner necessary to sustain it in a subsistent manner, while making possible the entire ionic form of life of the electromagnetic type of life, since this is the case with all ions everywhere, on Earth, in space, in fire, in the ionosphere of Earth, and in all stars.

The lively intelligent encoded oscillation of our proton in the electromagnetic field of the empty space is capable to form, hold, and maintain intelligence itself, while in this manner, it forms, holds, and maintains life itself, simultaneously. Intelligence is simultaneous with life and with interconnectivity, because our proton is not alone in space, but there are others, for each cubic centimeter of the empty space, always affecting each other in a continuous interconnected manner, while being capable to form together all the necessary inner realities capable to hold all specialized intelligences tending to the entire empty space between stars. Yet it only seems that life, intelligence, and interconnectivity take place simultaneously, because they are the same, only seen differently from different perspectives.

If life on Earth ever dies, which is impossible, then the ionic form of life of the empty space can always entrain and jumpstart the ionic form of life from Earth, which restarts life in all the life forms above in only a few million years, all the way up to the organic form of life and to the entire living ecosystem class of Earth. However, the ionic form of life from Earth, from space, and from all stars is in a continuous interconnectivity, and can never die, because its niche is always present here in this world and in the entire wider world, from the very low temperature of two kelvin of the empty space, to the ten thousand degrees of all stars, with the ionic form of life present everywhere and thriving, never dying for as long as the charged ions are capable to form electric and magnetic fields while oscillating continuously, because this is life. The oscillatory motion itself is not life, but only the lively encoded oscillatory motion, the code itself, only the lively intelligent code formed and maintained by this oscillatory motion of the charged ion in the electromagnetic field.

This marks the line between the living and the nonliving, this lively intelligent code of the oscillation, yet with life everywhere, all ions of this world and of the wider world will always oscillate only in a lively intelligent manner, and you cannot stop it, because life is the wider world itself, including the electromagnetic field, the stars around, the planets and all the ions, along with their continuous oscillation, always alive, always intelligent, and always interconnected.

It is similar with computers, because you also have two oscillatory motions to consider, one formed by the mechanical little switches of the motherboard in the real world from the close state to the open state and back again, and the other one formed by the zeroes and ones of the circuitry if electricity can pass or not through these switches. While again, this digital computer matrix is capable to form, hold, and maintain entire computer inner realities holding all software and all lines of code considered artificial intelligences, while all these intelligences know well how to maintain the entire computer software in a good functioning order.

As stated, with no life present, all charged protons of the empty space oscillate in a random manner. While with life around, they oscillate in a lively intelligent manner while making possible inner realities capable to hold intelligences in inner worlds correspondent with the outside world, while in this manner, our oscillating proton is alive.

Their oscillation of our protons is not free or random, since their motion is entrained in a forced oscillation by the neighboring protons. Their combined oscillatory motion as a whole forms electrostatic, electric, and magnetic fields, which is what I call the field. Life exists even in interstellar space, as it can propagate through and within this field through living modulated signals, which can be perceived as entrained, enforced intelligent oscillatory motion influenced by other protons, the field particles. This is the ionic form of life, part of the electromagnetic type of life present everywhere, in liquids, solids, gasses, space, in the plasma of fire, stars, ionospheres, and at the base of the molecular form of life, cellular form of life, and organic form of life.

Due to the low density of hydrogen ions in space, life is of very low resolution there, and therefore of a very low intelligence level. However, considering the immense vastness of space, as a whole, life is still relatively strong in space. It is of a lower existential resolution, but still more than enough to hold you there, or to transit you from one star or liquid medium to another, as you project or as you exist there permanently as an intelligence in the interstellar space. Since the electromagnetic field is compatible everywhere in the universe in the ionic form of life, in all its mediums.

Why exactly is intelligence of lower developmental level in space? Because space contains only ions, which are hydrogen atoms and hydrogen ions or protons, charged or not, depending on their atomic state. Since they can exist in several distinct states, depending on how these are affected by the electromagnetic radiation coming from stars. The field can change these atomic states at will, and this can increase the diversity of the field, making it capable to hold more life. These ions are capable to change only simple electromagnetic states of living modulation in the overall life held by the field, as they form, hold, and maintain the necessary ionic matrix holding all realities of the field.

The same electromagnetic field holds all intelligence in form of realities in all possible mediums: solid, liquid, air, space, and fire or plasma, which are very similar to the virtual world held by computers, and similar to the mind worlds held by living beings throughout all cognitive systems, because we find everywhere the same electromagnetic field, part of the same electromagnetic type of life. With the organic life included, if that particular medium is capable to offer the necessary niches holding organic life.

It seems that everything is an ionic form of life, because the ionic form of life seems to stand at the base of all forms of life, including the organic form of life. You can see it yourself, since when you study all organisms, you find them composed of the same ions and charged molecules oscillating in the field at the base of all cells and organs, making the organic form of life possible. Even cellular membranes and the nucleic acid oscillate continuously as plasma, since everything throughout cells and cellular components hold the cellular form of life and organic form of life. While at a closer study, you find the electromagnetic type of life with its ionic form of life at the base of the orbs, plasmoids, and rockoids, since everything is alive continuously everywhere in all worlds and realities of the wider world.

How exactly does life originate? It is always there, in the electromagnetic field. This is the case for the electromagnetic type of life, yet how exactly does the electromagnetic field originate? It is always there, right on top of the spacetime continuum, as it holds directly life, along with all natural laws of that particular world, making possible all lines of causality, lifelines of causality, and timelines of that particular world. Which is the same in the entire wider world, making Life herself possible everywhere continuously.

Furthermore, it is the same field holding computer worlds, and this is why, if you ever give your mind the means to learn to interact with computer worlds and computer software directly, it can do so, it can use its plasticity to interact directly with computers, increasing your cognitive reasoning including your learning and memory, while allowing you to network with other minds using technology, potentially with all minds, in an extraordinary collective consciousness.

These simple ions also exist in liquid, in water, in all oceans, and in all cells. Electric and electrostatic charge is not held only by hydrogen atoms in all mediums, but by all elements, radicals, and compounds, forming simple, distinct positive and negative charges, dipoles, and multi-poles. We have encountered ions of calcium, potassium, magnesium and sodium managing the sequence of patterns of electric charge distribution across the cellular membrane, including the interneuron synapses. This is one way for our cells to interact with the field, yet they can interact directly by changing their own overall field modulation, through all their ions and proteins counting in zillions.

I still distinguish between matter and the field, between particles and electric and magnetic field, yet at the very small scale of nanometers where we are now performing our research, there is not much difference. Because elementary particles, ions, radicals, chemical compounds, and molecules are nonsolid, shapeless, boundless, nothing but field, and part of the field. This is how the field can interact directly, intelligently with all matter, and this is how and where mind over matter is possible and therefore where reality is bent.

How exactly everything takes place in the field, lively and intelligently, while the field is rather dead and inert, as we study it in its two states, electric and magnetic? The field has many states, including its electric and magnetic states. We already know how these two states of the field are and how they make this entire world possible, since the electric and magnetic field stands at the base of all elementary particles, all molecules that these form, all solids, liquids, and gasses, and all matter and electromagnetic radiation in the world making all objects, subjects, communication, and technology possible including all industry, cybernetics, transportation, media, finance, and education.

Additionally, we have studied throughout this book and throughout this entire book series "Human," the specific intelligently modulated rapid fluctuation between the electric and magnetic states of the field, as it is controlled by life here and throughout the wider world, making life possible. This specific lively intelligently encoded field fluctuation between the magnetic and electric states of the field is not exactly material and objective, but it is abstractly intelligently encoded, and therefore it is out of this world altogether. It is an actual intelligent lively matrix capable to hold, form, and maintain all realities of the wider world, or at least those based on the electromagnetic field.

Our world is based on the electromagnetic field made possible in the common mind of all souls coming in this world by the billions. However, according to millions of testimonies, they use very advanced computer technology along with mind technology in our upper reality, in order to allow billions of souls to interconnect cognitively, and therefore in order to make our entire world possible. Yet this is the case not only here in this world, but in our entire cluster of created realities, since souls have souls that have souls that have souls in our entire cluster of created realities, with this particular world somewhere at the bottom, and with no other realities below besides the multitude of inner mind realities that you form right now while reading this book in order to reason intelligently in parallel with this book.

It is similar with computers, since computers form inner subjective digital matrices through the specific intelligently abstract encoded variations made by the switches of the motherboard, forming in this manner the intelligently encoded succession of the ones and zeros of the entire computer language, while this is the actual abstract digital computer matrix capable to hold, form, and maintain all computer software, and therefore all artificial computer intelligences roaming all inner, subjective, digital computer realities, including all videogame realities.

Computers are not alive, but only artificially alive, at the first mechanical algorithmic level, only that, as you study computers, you find them made possible by and through the same field making possible the entire ionic or plasmatic type of life, which includes the organic form of life, with the cellular form of life, molecular form of life, and ionic form of life at its base, with the field altogether at its base. Which makes all artificial computer intelligences compatible to all intelligences of the field, including the zillions of intelligences roaming the human organism.

Within our model of the organic life, we are exactly at the stage where we study everything in the liquid medium, exactly as it is made possible by the multitude of chemical and ionic reactions taking place in the liquid medium. However, it is the same ionic form of life here on Earth at the base of all living beings, right on top of the electromagnetic field as we found it in space formed by all oscillating protons of the empty space, only that within living beings, the ionic form of life is within all cells and cellular components, making them all alive and intelligent. With Life thriving continuously, off the significant energy exchanged continuously by the multitude of chemical and ionic reactions taking place in the liquid medium, all made possible in the field, through the field, and by the field, by the raw intelligences of the field more exactly, as these are held, formed, and maintained in their inner realities by all ions and molecules of the field throughout all these reactants found within all cells forming all organic life from below. To be even more precise, right now we study the chemosynthesis part of the organic life, exactly as it took place comprehensively over four billions years ago, before sunlight touched the surface of Earth. Because we had photosynthesis afterwards, along with chemosynthesis.

All these intelligently coordinated chemical and ionic reactions take place everywhere within the organic form of life, in all cells and therefore in all organisms, between atoms, ions, dipoles, multipoles, and molecules, which are animated by the same intelligences of the field, at the ionic and molecular levels, part of the ionic form of life and molecular form of life, standing in this manner at the base of the cellular form of life and further at the base of the organic form of life.

Which means that you cannot model the organic form of life apart from the cellular form of life, molecular form of life, and ionic form of life, because you have to model the entire electromagnetic type of life in order to understand the organic form of life very well, and this is what we have to do. We have already started with the field itself throughout the previous chapter, then we continued with ions and atoms in the paragraphs above, and now we continue with the actual chemical reactions taking place between these, within the liquid medium, standing at the base of the molecular form of life, and making it possible.

We also notice how we have to switch perspectives

continuously. Because we can study ions oscillating lively and intelligently in the electromagnetic field making possible the entire ionic form of life, while this is only the physical perspective of the entire electromagnetic type of life.

Oscillating ions in the field make possible all inner realities as these hold the necessary living intelligences capable to coordinate all ionic oscillations in the electromagnetic field along with all ionic chemical reactions in the electromagnetic field in a specialized manner, making all life possible at the ionic level, molecular level, cellular level, organic level, and social or ecosystem level. While this is a cognitive perspective, addressing intelligences themselves along with entire systems of intelligences and entire minds that these create, always in a specialized manner.

Furthermore, we notice how everything is mostly cognition and intelligences, while this is the supreme characteristic of Life: Mentalism. The continuous correspondence between the ions and proteins oscillating in the field and the inner realities made possible in this manner is the supreme characteristic of Life: Correspondence. The continuous oscillation itself at the base of life in the electromagnetic field is the supreme characteristic of Life: Vibration. While the specialized intelligences making everything possible in a specialized meaningful manner are the supreme characteristic of Life: Meaning. As these are part of our tenth level supreme perspective of Life, Intelligence, Interconnectivity, and the wider world.

There are more perspectives to consider, as the consensual, social, familial, and developmental perspectives, as we have always noticed them throughout the book and throughout the entire book series "Human." Because the same eating and digesting subconscious intelligence feeds you as an organism, feeds all your cells and cellular components, as it feeds your entire family at home if you are a mother, and the entire community at the restaurant if you are also a chef. The same intelligence continuously, your own subconscious eating intelligence, as it forms all the necessary systems of intelligences within your subconscious mind to make everything possible.

Yet it is the same intelligence, always found right on top of the electromagnetic field in the ionic form of life, since all intelligences are found in the ionic form of life, sitting right on top of the electromagnetic field, made possible by the continuous intelligently encoded oscillation of all ions of all living beings in the electromagnetic field. However, all intelligences of the ionic form of life interconnect in very large numbers to form all the necessary systems of intelligences necessary to govern all molecules of the molecular form of life that sits right on top of the ionic form of life, by meaning or specialization, while tending in an intelligent specialized manner to all specialized molecules of the molecular form of life.

Notice how from a cognitive perspective, you consider all intelligences and systems of intelligences, while from physical perspectives, these systems of intelligences are exactly the specialized proteins and enzymes of the molecular form of life.

However, all proteins, enzymes, RNA, ribosomes, and the rest of the molecules of the molecular form of life do not oscillate in the electromagnetic field of the cell individually, but they oscillate by the billions in a synchronous manner that unites our systems of intelligences even more, while making possible all cellular intelligences of the cellular form of life, also by specialization, in a meaningful specialized manner, tending to all specialized tasks making the entire cell possible, along with the entire cellular form of life. Similarly, all systems of intelligences of all cells unite to form even larger systems of intelligences spanning entire tissue, organs, bodily systems, and the entire organism if necessary, while even exiting the organism in order to tend to similar specialized tasks in the outside world, as feeding the entire family at home, the entire community at the restaurant, or even the entire nation, if this is the job that you have, through the same interconnected little intelligences of the ionic form of life made possible by the continuous oscillation of all ions and molecules in the electromagnetic field.

From a cognitive perspective, you can follow all these specialized intelligences and systems of intelligences in order to understand life. From physical perspectives, you can follow all cellular components with all ions and molecules included in order to understand life better. From living perspectives, you can study all living beings involved, as all cellular components, cells, organisms, families, and communities in order to understand life more, while you can even switch to chemical perspectives, because life makes possible the entire organic chemistry, made possible by all subcellular living beings through all their specialized systems of intelligences.

The ionic form of life makes possible all forms of life above, with the organic form of life included, because all intelligences live life in the ionic form of life, right on top of the electromagnetic field, interconnecting from there in very large numbers to make all cognitive systems possible, all living beings possible, along with their entire specialized interconnectivity and therefore the entire life on Earth. Offering you now even your meaning in life as a chef feeding your community, and you are fulfilled. With Life accepting you continuously with arms open, since you fulfill her in your own meaningful specialized manner, because your own subconscious eating intelligence feeds her through you, since it is the same. The same eating intelligence, having its roots in the raw field making everything possible. Which is part of our social perspective of life, highlighting the specialized meaning in life that all living beings and intelligences have, while making the entire wider world possible in a living intelligent meaningful specialized fulfilling manner.

In this entire living intelligent interconnected meaningful harmony, notice how all your intelligences are primarily the intelligences of Life, always fulfilling Life through you, more than they fulfill you. Your own intelligences will even kill you if you remain harmful, meaningless, and unfulfilling in life, by never tending to you anymore, or by producing free radicals in large numbers within your cells taking you apart from the inside out, and you get sick and die, if you remain meaningless and unfulfilling for life.

There are drugs currently keeping you pleased even while you are meaningless in life, yet you are pleased only artificially, temporarily, and with dreadful side effects, which is not exactly life. If you ever want to withdraw, start being meaningful in life and in the world in any manner, and you feel meaningful and fulfilled again, never seeking drugs again. However, the all intelligent human meanings are strictly controlled in the current consensual human society, and you are not allowed to fulfill them, while ruining the world in the process, because with billions of distinct very important intelligent human meanings remaining unfulfilled in life and in the world, this entire world decays to the first consensual tyrannical servitude level, while making all tyrants possible.

It is a vicious circle, because it is enough to take drugs in a consensual society to remain meaningless and unspecialized in life, make life impossible. While you are punished continuously by Life as you remain meaningless and unfulfilled in Life, since this is why you have to take drugs continuously in a consensual world. This is our consensual perspective in life.

It is no coincidence that all these perspectives show up together right here at the base of life in the electromagnetic field, since everything alive shows up exactly here at the base of everything. Because if we consider everything only from the physical perspective, it takes us nowhere, exactly as it takes the current biology nowhere, because the current biology studies life only from physical perspectives, while never being capable to understand life itself. Yes, the current scientific discipline studying life calling itself biology is incapable to understand life itself, because it never leaves this world in order to study life, while life always takes place at the confluence between two or more correspondent realities simultaneously, one physical and one cognitive. It is no coincidence that the current psychology is incapable to understand the human mind, because it never leaves the real world in its research, while the human mind is always found in an inner subjective cognitive mind reality, not in the real world.

They are looking in the wrong place the entire time, while confusing everybody in the world. Because if you have erroneous knowledge about life, reality, existence, cognition, humanity, meaning, and the wider world, you can never conduct accurate intelligent mental models throughout your entire reasoning, because your entire intelligent reasoning is based on accurate intelligent knowledge of life, reality, existence, cognition, humanity, meaning, and the wider world, making everything that you ever try to learn and elaborate inaccurate, useless and even harmful. No wonder everybody in the world takes drugs, because everybody in the world remains meaningless and unfulfilled through an erroneous reasoning, ruining themselves, their loved ones, and the entire world, one dark age after another, while these never end unless you are able to maintain an accurate successful fulfilling intelligent cognition, intelligent behavior, and intelligent interconnectivity within your mind and in the outside world.

As seen, all intelligences of the organic form of life and all intelligences of the social class of life are formed directly from below by the same raw intelligences of the field, as these interconnect lively throughout the ionic form of life to form wider intelligences and cognitive systems, interconnecting further within the molecular form of life as necessary by specialization to form wider systems of intelligences, which interconnect further throughout cells and throughout the entire cellular form of life as necessary to form wider cellular cognitive systems, which interconnect further to form even wider, organic cognitive systems spanning the entire organism, as it is the case with all bodily systems as the circulatory system, digestive system, reproductive system, and immune system. More precisely, whatever you see physically as a bodily system, from a cognitive, subjective perspective it is a wider specialized system of intelligences carrying their own specialized tasks within the entire organism, as feeding, breathing, reproduction, recovery, social behavior, and security. All our perspectives are always correspondent, since this is the supreme characteristic of Life, Correspondence, and you cannot have life otherwise.

As a reference, you know that bacteria are of the cellular form of life, while viruses are exactly of the molecular form of life. While when you study closely all viruses and all cellular components, you find them mostly as molecules. However, for the entire cell as a whole, we consider it of the cellular form of life. It is similar with organisms, since even if they are formed of cells, organisms are of the organic form of life. While you can always find large groups of molecules, as RNAs, proteins, ions, and enzymes living life always together, without a cellular membrane around them, and this is the molecular form of life. Viruses are an example, and you can find them in very large numbers.

Similarly, oscillating charged particles can create simple living intelligent modulation in the field from a physical perspective. Yet from a cognitive perspective, oscillating charged particles are simple thoughts or simple intelligences in the field, as these very weak ionic intelligences are barely capable to control their own living intelligent oscillation in the field as an entire oscillating ion in the field.

You need both the physical and the cognitive in order to understand life. You cannot divide the world only into the objective physical body, or only into the subjective mind or intelligence, since the two are correspondingly the same from comprehensive perspectives, and can even interact directly if you only adopt higher perspectives and higher abilities, to perceive the wider world as it actually is.

Are there enough elements and chemical compounds in the water, in our seas and oceans, to hold and maintain the entire electromagnetic field? Because our water is as pure, as clean, and clear as it can be. This is an achievement of life, since it already ate everything from all water. While back in the days when life morphed into the molecular form of life from the ionic form of life, the oceans were not too large, just appearing then, and not too clean, since there were volcanic vents everywhere, spitting elements and chemical compounds everywhere, as there was even more lava flowing in the oceans. The water was very hot, full of chemical compounds, reacting strongly everywhere, making a veritable nutritious soup, for all lithotrops out there to subsist, develop, and prosper.

Is life at the molecular level alive? Is life at the ionic level alive? Are molecules and ions alive? Yes, certainly, if they carry intelligences that have their roots in the field, since these give them life, through their source and essence, throughout their continuous interconnectivity. The ions and molecules of your computer hardware follow computer code and not raw intelligences of the field, and therefore they are not alive, but are only first level beings, caught in a first level algorithmic digital existence. However, as seen, life is everywhere, animating all particles and radiation of the field, and therefore all objects are animated by Life and Intelligence, at least at the ionic or molecular forms of life, and not necessarily at the cellular and organic forms of life.

Therefore, even the ions and molecules of your computer hardware are part of the overall ionic form of life and lower, and through your additional digital, algorithmic intelligences forming the computer software, you could easily be able to tap into the raw intelligences of the field already present within the molecules and ions of your computer hardware, if you only know how.

It is the same with the human mind, since through the ions and molecules of the ionic and molecular forms of life standing at the base of the organic form of life, you could easily exit the cognitive sphere of the organic life, to access the field directly, and therefore to project consciously anywhere you please. Animals and plants already do so, while humans are constrained directly to remain within their organic sphere of conscious interconnectivity. This happens at night anyway, when you sleep, and every time you lose consciousness, when the entire organism remains unavailable, and when you can exit the cognitive sphere of the organic life to wander wherever and whenever you please throughout the field.

What exactly are these intelligences inhabiting and roaming

the field? Why don't we just consider them simply part of life? Intelligences always exist in different realities and never in the real physical world where their physical bodies are. Intelligences are alive there in their own inner realities, behaving there as all normal living beings do. Life transcends through intelligences to inhabit simultaneously this world, their realities, along with all realities of the wider world, depending on the physical bodies with which they are associated. Intelligences can be considered living beings as they are, associated or not with physical bodies from this world or from other realities.

All living beings from this world have their own intelligences in their inner realities, forming in this manner entire living beings. However, there are living intelligences in the field nonrelated to this world, yet still benefiting implicitly from the interaction between the field and this world. While many times, they still influence this world themselves, implicitly. You can never find living beings lacking intelligence. These are called nonliving beings, objects, bodies, or corporations, from *corpore*, which means body in Latin.

Many of these intelligences are our intelligences, the intelligences of all living beings from this world, including all human, plant, and animal intelligences, from peptides and proteins up. Other intelligences are simple projections made by beings or by their intelligences from one reality to another, or from one medium to another, as your concepts, dreams, and astral projections. All these projections are perfect, segmented replicas of original intelligences, many fading away shortly, while others remaining alive for longer, or endlessly, depending on their strength and emerging awareness. Some intelligences are young, and others are very old. You know them, since you interact with them throughout your comprehensive fulfillment, learning, reasoning, dreams, and projections.

These intelligences manage our amino acids, peptides, and proteins, making them alive. Some intelligences fail, while others succeed and remain to manage communities of proteins, and continue as cellular consciousness. It is more than management, since it is pure existence, experiencing pure life, genuine experiences, learning, original knowledge, new abilities, possibility of ascension to new developmental levels, new forms of life, and higher status in the wider world. While all these make you stronger as an intelligence. If you fail, you can always go back and try again, since there are many beings born throughout realities, and you might claim anyone you want, if you only know how. However, know your limit, since if you take your chance or follow your greed exceeding your abilities, you are doomed to fail repeatedly, discouraging you, and even stopping you from developing.

Life in the field is far more complex than what I state here, since I do not have the time nor the patience to highlight all details. There are either harmony or competition associated with all life in the field, and no one remains unaffected, even here at the organism level. Everyone interacts, works, cooperates, or fights, yet as you study closely the cell, the human organism, the pasture, or the forest, you notice a continuous harmony everywhere, a peaceful fulfilling harmony, interconnecting rabbits with bacteria, proteins, lipids, earth worms, strawberries, stomachs, peaches, cats, and horses everywhere.

We know that throughout intuition, you always have memories and feelings coupled together. This happens in order to allow all cellular intelligences counting in zillions to help each other harmoniously throughout all specialized activities that each cellular component and each cellular intelligence has. Since even within cells, cellular intelligences send each other needs for all tasks that they cannot accomplish on their own, if it is not part of their own specialization.

It is very common for intelligences to send needs to other intelligences according to their specializations, by using feelings. While you can even feel your intelligences at work through all feelings from the background of your cognition. Yet if you feel only two feelings this entire time, pleasure and pain, it is enough to lower the volume of your pleasure in order to be able to feel everything present in your cognition, upfront and in the background.

What exactly do you do when no one can help you from among your entire organism, regardless of how many needs for help you send? You develop, since you have no other choice. More precisely, they develop, since they do not have a choice. Because you are always together, all intelligences of a cell, of an organism, of an entire community, of an entire species, or of an entire ecosystem, throughout an entire common life, and you have to maintain all inner and outer harmony instated. Otherwise, you end up one against another, you step on each other's toes deliberately or only accidentally, and this is how living beings die, genetic lines end, and species go extinct. While now you have to help each other to find all the necessary means to develop.

Organic life recycles solutions of old problems, adapting them for newer problems and projects, even by placing them on top of the old ones, in a rather artisanal manner. This might sound trivial, but it is not. Instead of starting new projects from scratch, when it seems more appropriate to us to dump everything and start all over again, since it would lead to more efficient results, organic life always uses and adapts its older solutions, since this technique of employing old solutions is at its roots, which is everything that it knows, since it is its most basic routine. This behavior is not as simplistic as it seems, since in time, organic life gathers zillions of solutions, capable to use them all, yet it will never step outside its baggage of solutions through pure invention of new solutions, since it cannot, because you can never know what you do not know. While as a cellular component, cell, organism, or entire community or society, you can never know to do what your own intelligences and living beings do not know what to do.

Therefore, however life was back then, billions of years ago in the ionic form of life, its most complex ions were the common amino acids that you still find currently, because these were formed of the most common chemical elements on Earth, and could not be otherwise. Additionally, their intelligences cannot die and are still around, they are currently at the base of the entire electromagnetic type of life with the organic form of life included, while these intelligences cannot be otherwise if you use different molecules at the base of your proteins, because you must use different ionic intelligences altogether, which is impossible, because these are at the base of all systems of intelligences and therefore at the base of all cognitive systems, while they are the only ones knowing what to do under all circumstances.

This is a problem, because it seems that the ionic form of life is already too rigid to change itself according to all changes in the environment while developing further. Yet the ionic form of life is too old and too complex to change anymore, with the laws of physics and chemistry never allowing its chemical compounds to be otherwise. This is why the ionic form of life made possible the entire molecular form of life on top, in order to surround it as a cushion in the continuously changing environment of Earth.

Being at the root of organic life, and at the very root of its routines, the amino acids used by organic life will never match with their structure and ability every change in the environment, but will always be used the way they are shaped currently, out of the same chains of reactants and chemical groups. This is why, once the environment changes drastically, when these amino acids are no longer adequate, the molecular form of life cannot change its chemical structure either, since the laws of physics and chemistry do not allow. Therefore, the organic form of life either perishes entirely, to return life to the cellular, molecular, or ionic forms of life, or it has to find other means, as morphing into a higher form of life altogether, if it can. Yet even if it decays to the ionic form of life, it can still ascend to a molecular form of life in the future, based on different main molecules this time, and not on the proteins and amino acids currently in use, but on whatever the new environment offers. While whatever the new environment can supply most abundantly, it becomes the norm, and nothing else.

Because you cannot simply tweak a protein within a cell,

and here you are, capable now to survive a major draught in the outer world. You cannot even change the chemical structure of a protein, since chemistry itself does not allow it at the very small scale of a protein. Yet since the major intelligences of your cell use consistently that particular amino acid or that particular protein, you cannot simply discard them, because you lose in the process your major cellular intelligences. What do you do? You have to develop, but you have to be very smart, since as you notice, at the very small scale of ions and molecules, the laws of physics and chemistry are very rigid.

These radical situations always occur whenever the environment changes drastically but you cannot change yourself to cope, mostly if you happen to measure in nanometers, with the entire chemistry against you. Since this is how life has to change entirely from one type of life to another with each drastic change from one environment to another, which is unfortunate.

Yet if your intelligences are very capable, you simply build a wall around your community, capable to keep the entire dreadful environment out. This is how molecules invented the cellular membrane, made the cell, and with it, made possible the entire cellular form of life, while keeping with it the entire electromagnetic type of life instated on Earth, which is an achievement for the ionic form of life.

Since this is why Life does not have an origin in any of its living realities, because even if it ever degrades to any of her lower forms of life, it can always ascend from there to any other upper form of life, according to all new major changes in the environment.

This happens if the basic elements and simple radicals mentioned above are no longer abundant on Earth, which is impossible, since they are always recycled from one compound to another and they cannot just disappear from Earth. Yet it can still happen, if the liquid medium itself, water, is no longer present on Earth, as it became depleted on Mars. However, even then, life is capable to carry its own environment with it, in its own bags, to survive even in dry environments, as in solid rock, deep underground, in dry deserts, or in solid, cold ice. While Mars did not lose its water, but it is still there, frozen underground. The organic form of life is very scarce in a very cold, frozen environment, yet the ionic form of life is still present, yet not as abundant as it is the case with the molecular, cellular, and organic forms of life on top, protecting it throughout all changes in the environment. While Mars became too cold to hold liquid water through dreadful natural or artificial circumstances if it had an atmosphere and it lost it, or through astronomical circumstances if the Sun aged gradually and became too cold to heat it accordingly.

Therefore, the molecular form of life is not the most efficient form of life anymore in a drastically changed environment, and it has to morph into a new form of life, the cellular form of life, which devours its niches now, and replaces it almost entirely. Yet since the cellular form of life sits right on top of the molecular form of life, it protects it as a cushion, and the molecular form of life with the ionic form of life within do not have to worry anymore of major changes in the outside environment, because they are both protected now by a cellular membrane, with the entire cellular cytoplasm maintaining the old environment of Earth almost intact, behind the thick cellular membrane or cellular wall. This is again an achievement for the ionic form of life, while as we start to notice, achievement characterizes life the most, and this is why I use this word in the definition of life.

Cells have cellular walls and cellular membranes capable to maintain the proper inner environment making the molecular form of life possible, within genuine cells of life, and it works very well. Because if the molecular form of life could modify its structure so easily, it morphed its molecules according to the new environmental conditions, and kept living normally throughout the oceans and seas. Yet it cannot, through very rigid physical and chemical properties at the molecular level, as bounding requirements and ionic conservation.

Which means that the current twenty amino acids used

throughout all proteins of organic life have to stay, because they cannot be modified, while if the environment changes drastically making your twenty amino acids impossible on Earth, you either move away, remain alive only in the very few pockets on Earth as in all volcanic vents, or you simply build a wall around you in order to keep the toxic environment out, making the entire cellular form of life possible right on top of the molecular form of life, because the molecular form of life can continue to live undisturbed only within cells.

This is how Life transcends to higher forms of life, through all drastic changes in the environment. Yet when you study closely the particular drastic change in the environment making possible the entire cellular form of life, the oxygen event, you notice how it was made possible by life itself, by all molecules seeking to use sunlight as source of energy instead of chemosynthesis, switching to photosynthesis instead. Yet sunlight is a major source of energy on Earth and cannot be ignored, and this is why life preferred to create an entire new form of life, the cellular form of life, right on top of the molecular form of life and of the ionic form of life, only to take advantage of sunlight. It was better if life could modify its twenty amino acids in order to survive the oxygen event in the open, which was significantly easier, yet it was impossible by the laws of chemistry, since oxygen is very corrosive. While if it abandoned its twenty amino acids, it lost most of its molecular intelligences, and could not survive without them, because intelligences are life altogether. More precisely, if you have to change most of your inner intelligences at the base of life, you must change life altogether, which becomes an option only with the old life dead. Yet since the old life survived the oxygen event, by building an entire cellular membrane and therefore by making possible an entire form of life sitting right on top of the first two, the old life is still around now, not a new one.

The molecular form of life will still exist freely outside cells, but only within niches inaccessible to upper forms of life, as we still have currently all the previous forms of life occupying freely all remote niches inaccessible to organic life, as volcanic lakes and volcanic vents. Furthermore, the ionic and molecular forms of life will always exist at the base of the cellular and organic forms of life, within the cellular and organic forms of life, as you always notice.

Life adapts, improves, diversifies and expands to span the Earth, as much as resources allow, and it diversifies furthermore, by maximizing efficiency. Again, it is not only the natural environment changing and pushing the organic life to develop to its next level, but the continuous lack of resources, generating a tough situation, since life is everywhere, in all forms and through all possible species.

Life recycles older solutions for newer problems throughout its entire development, even when these new problems are encountered in higher or lower forms of life, as though it is the same thing, because the same intelligences live life in all forms of life, only in larger or smaller cognitive systems.

Therefore, first we have ions oscillating intelligently in the field while interconnecting with the field in an intelligent manner. As groups or ionic arrays, these ions can modulate intelligently the field, forming the right matrix to hold inner realities filled with intelligences that are actually the ones modulating the field intelligently, by coordinating the movement of all ions in a living, intelligent manner. Once you have a body and a system of intelligences, you have life, as this is the ionic form of life. However, we cannot know if there are even lower forms of life at the base of the ionic form of life, as atomic, nuclear, or subnuclear forms of life, but we only speculate that the formation of protons in the large clouds of protons from the interstellar space can be a living process, undergone by forms of life found at the base of the ionic form of life. Yet the ionic form of life itself should be able to create all hydrogen atoms from the interstellar and intergalactic space, by converting the entire electromagnetic radiation coming from all stars simultaneously, in an interfering manner.

Under specific circumstances, ions and elements in general

gather to form large molecules, through what science calls organic chemistry, distinguishing it from inorganic chemistry. These should be called living chemistry and nonliving chemistry, since it is life coordinating intelligently, in a living manner, all chemical reactions of the organic chemistry.

Therefore, through living intelligences, ions gather to form more complex chemical compounds called molecules, of the molecular form of life. While these molecules are always made according to the environment, according to everything that the environment can offer more abundantly.

More precisely, the ionic form of life makes possible an entire form of life on top of it, as an extension, in order to be able to reach all niches of this world. Life lived in molecular communities is a higher form of life, of the second form of life, called the molecular form of life. This is the physical perspective, while from cognitive perspectives, you notice how the intelligences of the ionic form of life gather in large numbers by specialization in order to form larger systems of intelligences capable to reach all niches of this world. The molecules themselves are these systems of intelligences of the ionic form of life, now living life in entire systems or communities of intelligences having the physical shape of molecules. Yet all intelligences of the ionic form of life cannot gather randomly while forming their systems of intelligences, because they must abide to the laws of chemistry. This is why there are not too many types of molecules forming the molecular form of life besides the proteins, enzymes, fatty acids, and DNA molecules that you know well.

It is the same with humanity, because humans do not live life alone but in entire families, communities, societies, and civilizations, while these are an entire living class of life sitting right on top of the living human beings of the organic form of life. From higher perspectives, there is no difference between forms of life and classes of life, only that you tend to refer to everything below you as a form of life, and to everything above you as a class of life.

Viruses are an example of a molecular form of life, along

with all communities of molecules as communities of proteins, enzymes, and RNAs living life together without a cellular membrane around them, and you can find them whenever the old environment of Earth is still present in restraint pockets, as in volcanic lakes and vents, and even in the digestive system of all organic living beings.

The third form of life is the cellular form of life, or the prokaryotic cellular form of life more precisely. Which comes to be this way through the change of the environment of the molecular form of life. When the environment of this world changed drastically billions of years ago, in order to survive, communities of molecules had to build walls or membranes around themselves, only to maintain a proper, favorable environment within, since molecules cannot remain stable in other environments, corroding or combusting relatively fast.

However, your own family at home and your own community where you live act as a cushion to the outside environment, maintaining within the community a milder environment. Within communities, the environment never changes as drastically as it does in the outside world, and as long as all members perform their specialized meanings right, there is a relative stability, harmonious interconnectivity, and therefore abundance within the community of ions, molecules, cells, or organisms. Yet stability and abundance cause members to develop less, yet their communities develop as a whole, as it interconnects with the entire environment, including with other communities. While at a closer study, you notice how the living members themselves develop, because they are the community itself, only that they develop more in a specialized organized manner, and less in an individual manner.

How does the community develop? It maximizes the efficiency within, it builds walls, it interconnects with other communities, but mostly, it maximizes the efficiency of all its specialized meanings, as maintaining, feeding, transporting, defeating, securing, improving, reproducing, developing, and diversifying everything. While at a closer study, you notice how all specialized meanings and tasks of all members from all communities of all forms of life and classes of life are similar, because they are the same, made possible by the same specialized inner intelligences of the ionic form of life, since they are the ones alive the entire time, throughout all higher forms of life and classes of life that they ever instate, maintain, and develop. All your intelligences are systems of intelligences, while they are all formed through the continuous specialized interconnectivity of all intelligences of the ionic form of life. You have only ionic intelligences within your cognitive system by the zillions, yet they are continuously interconnected in a meaningful, specialized manner.

As stated, all intelligences of the ionic form of life cannot interconnect tightly in any possible manner, but they must follow the laws of chemistry and the laws of physics affecting the compounded physical bodies that they form together throughout their interconnectivity. While in this manner, the intelligences of the ionic form of life do not have too many choices to interconnect besides the common molecules that you find within cells.

Proteins are made of hundreds of amino acids, or of thousands of amino acids, yet the amino acids forming them are of only twenty-two types. The amino acids themselves are of the ionic form of life, while the proteins that they form are of the molecular form of life. The intelligences of the ionic form of life interconnect tightly into proteins, which are always specialized, while remaining as proteins for as long as the specific specialized task is needed in the community.

The intelligences of the ionic form of life do not have to interconnect only tightly into molecules, but they can interconnect in a specialized manner from a distance, by acting and reacting together on the electromagnetic field in an organized manner for all possible specialized purposes, as moving themselves and anything else around, producing charge or energy, bringing in resources, replicating their physical bodies in larger or smaller numbers, or taking the waste out of the community. While their community where they live is under a continuous very strong electromagnetic field, making their specialized tasks harder or easier to fulfill, depending on circumstances.

Study proteins closely, to notice specific coiled antennas of emission-reception specifically designed to interact directly with the electromagnetic field, called helices, used to glide around in the electromagnetic field. Proteins can generate their own field using helices, along with other internal electronic structures resembling dipoles, capacitors, chips, and linear motors, used for propulsion, power generation, thinking, communication, attachment, towing, tools, and weapons.

All ionic intelligences of the ionic form of life make these possible, in order to act and react on the electromagnetic field in a lively organized meaningful specialized intelligent manner, while making some people assume that only an intelligent creator can make all these possible, because all ions and molecules of the electromagnetic type of life are considered lifeless and therefore unintelligent particles, exactly as the current science depicts them.

Ions, amino acids, proteins, and enzymes are exactly our nanobots from the previous model, as they interact directly on the field physically, without the use of a computer, yet they do so in a living intelligent manner, since they are the ones giving life to the entire cell and then to the entire organism. While if you consider the ions and molecules of the ionic and molecular forms of life automatic and nonliving, you must consider all cells, organisms, families, and communities nonliving, unintelligent, and automatic, because all life comes directly from ions, from their continuous lively intelligent encoded oscillation in the electromagnetic field making everything possible.

Ions are living beings of the first form of life, ionic life. Proteins and their entire community are living beings of the second form of life, the molecular form of life. Their living, enclosed community, the cell, including the cellular intelligence or consciousness, is a living being of the third form of life, the prokaryotic cellular form of life. There is a fourth form of life, the eukaryotes that we will study shortly, while the fifth form of life or organization is the organism itself, or the organic form of life. However, the current biology considers only organisms alive, but not the eukaryotes composing them, because biology persists to study only organic life and to consider only the organic life alive, but not the other forms of life standing at its base.

Are proteins alive? Proteins supply, tend to, and maintain their own intelligence throughout life, and therefore they are alive. Part of their intelligence lives within the reality spanning the matrix formed throughout the chemical and ionic bounding generating its own distinct field. The other part of its intelligence lives in the cellular field, and furthermore, in the comprehensive electromagnetic field of the entire environment. Proteins act directly, in a physical manner of the field, interacting with its entire intelligence. The human mind is conscious and subconscious, with the subconscious spanning the entire organism. The human intelligence is not entirely in the brain, but throughout the entire body. Furthermore, a slight part of our overall intelligence lives throughout the bodies of those close to us, and throughout all mediums around us making the overall field.

Proteins also build other proteins by copying them objectively. You can simply pick up a protein floating around, stretch it up in a chain, and start copying it from scratch amino acid by amino acid, ending up with its perfect replica that holds a different intelligence, or the same overall intelligence, whatever the field or the other intelligences assign. However, ribosomes can always shake newly created proteins in the surrounding electromagnetic field strongly enough, and intelligently enough, to force the newly formed protein to develop and to maintain exactly the specific specialized intelligence needed within the community. This is why you cannot take intelligences from other communities, cells, or organisms to use them as your own, because they are not the protein intelligences that you need, and therefore you must digest them or you must take them apart into the amino acids of the ionic form of life, in order to assemble them within your community exactly as you need them, with the intelligences that you need.

Yet proteins are copied mostly from the RNA or the DNA, which are filled with all the necessary prototype proteins that you should ever need throughout life. Because living communities do not exactly store abstract knowledge and abstract successful solutions of how to perform all the previous specialized tasks ever encountered since the dawn of life, they do not actually store subjective knowledge from one community to another, from one cell to another, or from one generation to another, but they store living intelligence prototypes, including their specialized intelligence having all the necessary knowledge and abilities to perform that specific specialized task. In this manner, specialized enzymes and specialized proteins replicate or bring to life the necessary number of proteins to perform the necessary task, and these perform the necessary specialized task within the community, within the cell, and within the entire organism, for as long as they maintain integrity, and therefore for as long as they are alive. If their task is no longer needed without notice, then these specialized proteins are disassembled immediately, recycled, and they cease to exist, stopping working in this blunt manner, while also losing their intelligences. Yet similar intelligences are everywhere within the community and in the nucleic acid stored as prototypes, and therefore the intelligent type is not lost.

It certainly makes a difference if you have an RNA within your community to store all your prototype intelligences, because without an RNA or DNA, you must store all your types of specialized intelligences randomly throughout the cytoplasm of your community, having to use these as prototype intelligences, while they might be already damages. By copying random proteins floating around, you also copy all their structural damages, since this is what happens when you make copies of copies of copies. Ideally is to copy an undamaged protein, a brand-new one. This means that you can save a protein of each type, only to have a model for replication, called prototype.

However, with thousands or hundreds of thousands of different types of proteins, you have millions of prototypes floating around doing nothing and being damaged anyway. What do you do? You store the prototype intelligences systematically. You simply take all prototypes and snap them on each other to form a long molecule of proteins. Each protein prototype within this long chain of proteins is called a gene. This entire molecule containing all prototypes, or all genes, is called an RNA molecule. While in order to preserve this entire chain of prototype molecules even better, you attach the head to the tail in a large circle, exactly as you find these within all prokaryotes. Yet you must always use specialized enzymes continuously in order to check the integrity of the entire RNA sequence, and then you must use other specialized enzymes to repair it if it is damaged.

All these are alive, including the entire RNA chain of prototype molecules, with all their specialized intelligences alive even while they are packed, while always living a normal specialized life either packed or unpacked. You can always tell by their specific continuous nonrandom lively intelligent modulated oscillation in the electromagnetic field, making the entire nucleic acid similar to plasma, because it is alive.

RNA molecules can function on their own, as gigantic living molecules, while providing all the prototype proteins that it contains every time they are needed. However, all prototypes within the RNA, all these genes are not collapsed into themselves as genuine proteins, but they are simply linear chains of amino acids, and therefore they do not help too much to create a living being out of the RNA itself. It is true that the RNA molecule is made of more than genes and instructions on how to create proteins using genes, and this gives the RNA molecule a relative autonomy within their living community.

You can still find currently the molecular form of life living in these large communities formed of enzymes, proteins, amino acids, and ions gathered around gigantic RNA molecules, with no cellular membranes around them, and therefore living life not in the form of a cell, not as a cellular form of life, but as a molecular form of life, exactly as all life lived before the cellular form of life, yet the current biology never studies them. Even viruses are of the molecular form of life, not of the organic form of life, since viruses are not eukaryotes. The current biology still considers viruses cells, which is erroneous, while considering them living beings or not, in a major debate. While viruses are alive, but not of the organic form of life, and not even of the cellular form of life, but of the molecular form of life, with the same ionic form of life at their base, and with the same ionic intelligences of the ionic form of life making them possible, by interconnecting continuously in the same proteins, enzymes, and RNA that you know well, making all viruses possible.

The prototype intelligences within the RNA, or the genes, are still damaged in time, generating short-lived and faulty replications, affecting the entire community, and must be avoided. Specialized enzymes check the RNA constantly for genetic damage, and repair it. How can enzymes repair the RNA? How do they identify the damages in the sequences of amino acids within the genes? The repairing proteins or enzymes must check the integrity of the RNA by comparing it with another RNA, for perfect resemblance. Once a lack of similarity between the two is found, the repairing proteins repair the RNA.

But which RNA? How can a protein tell which one is damaged and which one is not? It cannot, or at least not in this mechanical manner as stated by this simplistic model of organic life, but in an intelligent manner, through specialized protein intelligences that have verified and repaired RNA molecules since the dawn of the molecular form of life. Because as you notice, their consciousness never dies, but it is transferred from one community to another, from one cell to another, from one organism to another, from one species to another, and from one form of life to another, while all specialized intelligences are always the same. How old are you?

You are as old as this entire world, according to your intelligences.

The electromagnetic field is very intense at the small scale of nanometers of the molecular form of life, and therefore your existence within the field as a protein or protein intelligence at that small scale is more as a game of Puzzle, where you must fit perfectly in the tight, very strong field, or you cannot exist. Because the laws of physics and chemistry do not allow you, getting in the way of everybody, so they have to recycle you and terminate your existence. This is how all damages within the community at all levels can become major anomalies in the overall field, and everybody is eager to solve them while restore homeostasis.

Many times, this is how specialized intelligences perform their specialized tasks, by seeking or maintaining equilibrium, perfect interconnectivity, and maximum prosperity and efficiency. While whatever you cannot do yourself, you ask for help from other specialized proteins by sending them needs and by using the general system of punishment and rewards acting through feelings. Everybody does so, since this is how everybody interconnects within the community, in a specialized manner, by needs, fulfillment, rewards, and punishment, and it always works.

Yet it does not work anymore throughout human communities, because humans take drugs, while drugs interfere drastically with the entire punishment reward living mechanism, compromising the entire human interconnectivity, while ruining an entire world. Consequently, the entire living intelligent human social class of life is disabled currently, as it is the case throughout all dark ages, while making all tyrants possible. This is why humanity is addicted and consensual, in order to make all tyrants possible.

Maintaining the genetic integrity is always a problem and a main priority for living beings. Genetic integrity is maintained in various manners. First, within the RNA coding, the same genes are stored in multiple places in different positions, upside down and in parallel chains, while maintenance enzymes check and compare homologue genes, and they repair the damaged ones.

Why placing within RNA and DNA gene sequences upside down and left-right reversed? Because algorithmic processes can miss repetitive damages. When you place similar chains of genes directly in sequence, you risk ignoring their identical damages unknowingly, since you find them similarly valid, even when they are similarly damaged, always assuming that they are accurate, while they are altered, fooling you. Later on, all your recycling proteins disintegrate away in a matter of seconds, even the newly made proteins, because they are already damaged and ineffective, or even harmful. You order more proteins to be built, in vain. You do not know what is wrong, soon your entire community is but a blob of waste material, and this is how you die.

Yet organic life has safety checks on top of safety checks when it comes to prototype intelligences. There are specialized proteins checking all RNA in the community for the similar damages, and even other RNA molecules from other communities, since proteins can read the field of each RNA from a distance, through the common induction of the very strong field at their very small scale. Yet molecular communities and cells always send their proteins and enzymes to their neighbors as a reference for them to check for errors, and it works.

Because every time you find the nucleic acid entirely unwound, oscillating everywhere, taking all space, while enlarging significantly the entire cellular nucleus, you do not have cancer, but major maintenance takes place within the cellular nucleus, checking for errors, repairing errors, or modifying, upgrading, and developing the prototype intelligences themselves.

Proteins are alive and intelligent by all standards. Proteins replicate themselves, they use electricity, they glide throughout the field as maglev trains, they have minds, and they have electronic components. Proteins are specialized for each task within their community and they do everything with accuracy, always floating around to take care of everything, transporting and packaging resources, cleaning up, and always assembling and disassembling things. Proteins can also assemble themselves into very large structures as tubes, walls, poles, carts, cranes, rafts, and conduits. Yet their entire living community looks more as a larger house, as a factory, as a city, or as a civilization, through its structure, components, members, results, and overall activity.

How exactly does such an impressive community function? Can the cellular consciousness keep track individually of every little protein, making sure that they do everything that needs to be done, while checking personally every product that they make and every spot that they clean? No. The cellular consciousness only instructs how many proteins of a certain type are needed at a certain time in the community to perform a specific task. Once a protein is made and let loose in the community, it will perform its job accurately for minutes and hours at a time, as long as it lives. The community will not build one protein at a time, but batches of millions or billions of similarly specialized proteins at a time. Does the community need more energy? It builds a certain number of proteins specialized in the gathering of reactants and in causing and maintaining chemical reactions. Does the community need more charge? Specialized proteins are assembled to produce the charge through specific chemical reactions from specific resources. Is the field getting weaker? More specialized proteins are made to maintain the field by injecting charge wherever it is needed. In this manner, the community only produces more specialized proteins every time it needs anything to be done, and these proteins know exactly what to do and how to do everything, since their own specialized intelligences have always done so throughout the molecular form of life and long before, throughout the ionic form of life and long before, and now they know everything well.

Proteins live for seconds, minutes, hours, or days, depending on their type. RNA lives longer, yet the community of proteins can live endlessly, if it manages to survive, subsist, and develop, and if its environment allows it. Will the community divide, replicate or give birth to other communities of proteins? The community only produces more RNA molecules, as plasmids, or RNA circular in shape, and these live on, with the necessary living molecules always around them, as it is the case with viruses and with other molecular living beings. Yet usually, proteins swarm around their RNA, and this is how they live, always replicating themselves as necessary by using the RNA. RNA molecules also travel around the environment following sources of food, which is made of specific reactants, taking their proteins with them. Some of these communities formed of an RNA and a multitude of proteins and ions around are basic viruses, as we find them everywhere within and outside cells, since these are of the molecular form of life.

Proteins, through their collective field, hold their consciousness, and therefore the overall community intelligence follows the RNA molecule along, with all proteins around, wherever they go. This is a specific form of reproduction, done through colonization, which is found in almost all forms of life, including the human social class of life.

What do proteins eat? Proteins eat charge or energy, directly from the chemical reactions that they control either around themselves or within themselves. More developed communities produce and maintain the chemical reactions in a confined environment, and then they transfer rations of energy, charge, and electrons to all proteins of their community, even in form of ATP.

We are almost one billion years from the start of cellular life, which was over three billion years ago, according to science. These communities of proteins swarmed the world then, with more RNA molecules being formed and departing to colonize everywhere, throughout an incredible harmony or competition, depending on modes of life, modes of intelligence, and modes of interconnectivity. Currently, most colonies or communities of proteins are lithotrops, eating directly chemical compounds as calcium carbonate or sulfuric acid, along with pure elements as iron and sodium. Other communities do not go through the entire hassle of producing and controlling chemical reactions, but they prefer to capture proteins of other communities, take them apart, and use their amino acids and energy to replicate their own proteins and to fuel them with energy. While other communities feed on these communities that feed on lithotrophs, in veritable food chains. There are still proteins and RNA molecules currently living their lives individually and not as part of communities, either by replicating themselves occasionally, or by integrating themselves into the RNA genetic sequence from nearby communities, if they make it through undetected, as viruses do within cells. Their kind is replicated millions of times by others, to roam the waters freely and solitarily. Some dying of old age hours and days later, with very few to make it into the genetic code of other RNA, where they are replicated again, into millions of copies or more. These solitary proteins and RNA are still around, yet more diversified now, and we call them viruses.

The question is how far you can spread within your liquid environment when every micrometer cube of water is already taken, with all resources strictly managed to the last molecule, and with zillions of extra communities formed by the second? You can go out of the water, if you can, but there is no water out of the water, while chemical reactions can take place only in liquid. Yet the dry land of Earth is full of tasty, abundant chemical food, ready to be eaten, if you only know how to reach it.

Communities are forced now to go out of their native medium, out of water, and spread the Earth, entirely. How do they do so? Adaptation to the environment is the answer coming from the theory of evolution. However, we have already stated that all amino acids are highly stable and will never adapt or change their structure under any change of the environment, and therefore they cannot adapt and cannot develop, since the laws of physics and chemistry forbid them, being already at their most efficient form. Furthermore, chemical reactions take place in liquids the most, and not directly among solids. Therefore, molecular forms of life will never go out of the water. Water is the liquid medium of the molecular form of life, and will always be. In order to go out of the water to inhabit the land, the molecular form of life must morph again into other forms of life.

Are there abundant resources out of reach? Life can find a way to get to them. Are there entire vacant niches available on land, not too far away? Life will certainly find a way to occupy them, since Life had placed them there in the first place, because the entire environment is alive.

This is how molecular life gathered its RNA, proteins, communities, and stocks of resources, along with its entire environment, water, it packed everything up into bags of water, and went out of the oceans and seas to inhabit the land. The cellular intelligences created and instructed specialized proteins to build an impermeable membrane around the community, able to hold them, along with their medium, water. This is the cell, the third level cellular form of life. This seems tedious and hard to put together, yet our spaceships are nothing but cells, containing people, supplies, computers, water, and air. Everything happened on shores, where molecular communities wandered closer and closer to air, and found themselves out of water. For a short period at first, then they found ways to survive longer out of water, by using impermeable membranes of fats, and then they simply built permanent membranes around the entire molecular community, and this is the cellular form of life.

Yet cells did not appear only to be able to go out of the water, but in order to survive in all unfavorable environments, including poisonous liquid environments. Molecular communities had to build membranes and walls around their communities only to be able to preserve their initial environment, while the entire ocean environment changed continuously, favorably and unfavorably. They had to preserve their initial liquid medium, because as seen, molecules cannot adapt too much to their environment, and therefore they had to adapt the environment to them, keeping it favorable continuously within, as humans do. If they could not, they had to preserve it in any manner, by building a membrane around it, keeping it unaltered in this manner.

Because the composition of oceans changed drastically repeatedly throughout time, as it happened when the process of photosynthesis poisoned the entire environment with excessive oxygen the first time when sunlight touched the surface of Earth. Which happened when the atmosphere of Earth cleared of smoke and volcanic ash for the first time, when the oceans filled up with waste, when they cleared up of useful resources, or now when molecular living beings had to go out of the water to expand its niche, and simply took the water with them, within and around themselves, as cells, while making possible an entire form of life, the cellular form of life.

Science never reports finding communities of proteins roaming around without a cellular membrane. Furthermore, science insists on stating that organic life had first appeared directly in form of a complete cell, including proteins, nucleic acid, and cellular membrane, even by chance. While molecular life appeared first with amino acids, and continued diversifying its complexity through peptides, proteins, and RNA. While now, it ascends to the cellular form of life, through the addition of the cellular membrane. While many viruses are still around, as a proof of those old times, when all life lived only in form of molecules. Proteins and RNAs float freely everywhere, yet they are always associated with cells and organisms, and never considered individually. Old communities of proteins still exist currently, not in our environment, but in pockets of the old environment of four billion years ago, still found currently in volcanic vents and around sulfur or phosphate deposits.

Cellular membranes are first formed in order to hold communities of molecules as proteins, enzymes, and RNAs safely within an unaltered liquid environment, yet there is more to consider, because cellular membranes are the brain of the cell, holding the conscious intelligence that can become pertinent enough to coordinate the entire conscious interconnectivity of the entire cell with its outside environment while it fulfills its outer needs, similar to your needs in the outside world.

Which means that cellular membranes are formed through the unification of all molecules specialized in the interaction of the entire molecular community with the outside world, as the security molecules, resource-gathering molecules, waste removal molecules, perception molecules, along with all molecules used for thinking throughout all cognitive processes involving the outside world. All these molecules specialized in the continuous interaction with the outside world always stand at the periphery of the molecular community, where the outside world starts, and now through their increased number, they unify entirely, forming the cellular membrane.

This is not a tedious task, since there is only field at that very small scale, and not too much rigidity involved. Furthermore, from the cognitive perspective, the inner intelligences of all these molecules unite to form the cellular membrane, only to allow them to confine and therefore protect the entire community.

Cellular membranes can keep all cellular components safely within, they can keep all resources safely within, while keeping all waste outside, maintaining a favorable environment within, maintaining a favorable environment unchanged indefinitely, and these are good reasons to have a cellular membrane, even if you have to make possible an entire higher form of life to have it. As seen, there are still ionic and molecular forms of life still around, capable to interconnect successfully continuously with cells and with entire organisms, occupying in this manner similar niches. Cellular membranes can define the boundaries of a community of peptides, so its members and resources do not get lost, they keep intruders out, they keep germs out, they shield the field and they keep it within the community, and in this manner, they create a designated larger conscious brain able to hold the cellular consciousness in and around the cellular membrane.

Why is the electromagnetic field not capable to create fully-grown organisms instantly, out of anything? It takes very advanced intelligences to be able to create everything out of nothing, since throughout life, everything is developed gradually, whenever needed, according to the environment.

Yet since throughout the entire gradual development of life, life preserves all specialized intelligences ever used, in case that they are needed again, it means that you can study the RNA, since the prototype molecular intelligences making everything possible on the way are still there, saved as genes. While you can find them in a modified form, as they are capable now to form related cellular and organic structures similar with the cellular membrane, as skin and brains. It seems that these specific specialized intelligences are capable not only to form similar organic structures if the environment ever requires and allows, but they are capable to shift life entirely to higher forms of life.

Cellular membranes strengthened communities, and therefore all communities adopted them even in the liquid medium, simply by direct communication between cooperating cellular consciousness, and by the simple, direct integration of the specific genes, the prototypes of the specialized proteins constructing the cellular membrane, maintaining it, managing the transport across it, and allowing the formation and succession of patterns of charge distribution enhancing cellular conscious thinking and intercellular communication.

We forward the model as Earth cools down slowly. The overall tectonic activity slows down and there is more water in the seas and oceans, part of it coming from the condensation of atmospheric vapors in what we call currently acid rain, what it was then very acidic rain, full of nutrients for all molecular and cellular communities, as all prokaryotes, including archea.

The Earth was still heavily clouded, covered fully in dust and ash, in an everlasting night. However, with less and less volcanoes, the atmosphere cleared out, slowly, and sunlight appeared, very dim at first, yet more intense throughout the centuries, until it was intense enough to be scavenged or harvested by Life as a very new form of energy.

A little over two billion years ago, a major event changed drastically life, making it resemble to what we have currently, aerobic life. Let us study it, since it leads us to higher forms of life, to eukaryotes, and then to organisms, to us.

The cellular and organic forms of life are very strong forms of life inhabiting the liquid medium, scavenging and feeding on the energy, charge, compounds, ions, and electrons generated by chemical reactions. Chemosynthesis. We have seen that the cellular form of life had found means of living in other mediums, as the air medium and solid medium, by carrying its own liquid medium everywhere. Now we are going to see how the cellular form of life is ready to abandon chemical reactions as a source of charge, electrons, and energy, using directly electromagnetic radiation from the Sun. It is typical for Life to intertwine and inhabit several mediums and realities simultaneously.

How can you harvest the energy of light? With solar panels, since this is what the thylakoid is, a natural solar panel. Yet what the thylakoid can do additionally to our solar panels, it stores energy into nutrients as food, it powers up little explosive batteries called ATP, zillions of them, sending them around the cell to power up all proteins, and it even coins amino acids out of elements and compounds with it. Life overpasses human technology, again. How does it do everything? By using the same solar technology, only better, and more efficiently. The thylakoid uses solar energy to break down water molecules through basic electrolysis, it harvests electrons, hydrogen, and oxygen in this manner, and it uses the energy to build carbohydrates out of carbon, oxygen, nitrogen, and hydrogen, carbohydrates that are food and storage of energy. We use them as food, while it uses the rest of the energy to coin ATP molecules and amino acids.

Science states that the first prokaryotes to develop thylakoids are the cyanobacteria. This new form of energy, solar radiation, is so efficient, that cyanobacteria had expanded shortly to dominate Earth. Cyanobacteria are still present currently, after two billion years of existence, living independently as cyanobacteria, living within communities of higher forms of life as algae, or living even within the eukaryotic cells that form organisms as chloroplast. Cells within cells! Yes, the prokaryotic cells chloroplast and mitochondrion live by the dozens within the eukaryotic cells that form all organisms. The eukaryotic cells might resemble prokaryotic cells, but they are twenty times larger or more, and I consider them a separate higher form of life, right below the organic form of life.

At that time, the massive amounts of oxygen that cyanobacteria generated changed drastically the composition of water and air, eradicating our protein communities, along with most of prokaryotes of that time. Oxygen is very corrosive, reacting strongly not only with carbon, but also with metals and other elements and compounds. All niches and all food chains changed completely during the oxygen event, forming new ones, based primarily on oxygen, for the aerobic life. With one niche based exclusively on the reaction between oxygen and carbon to form carbohydrates, defining animals.

How did the oxygen event manage to reshape life? The change took place at the third, prokaryotic form of life, reshaping drastically communities as a whole. All members of molecular communities, all amino acids, proteins, and nucleic acid lived and developed normally, unchanged. Yet this entire sudden change happened for one reason, because the molecular form of life can never be changed from its roots, since it cannot change the chemical structure of its amino acids, along with its basic molecular routines. This is the reason why life had to reshape the behavior of its communities entirely, forming cellular membranes, only to be able to maintain intact the environment within cells, as it was the case over four billion years ago, when water and molecular life had first appeared on Earth, when amino acids were first formed.

Because these amino acids are not stable anymore in a liquid medium of high concentration of oxygen, since they corrode, they combust, they disintegrate, and they lose consistency, disintegrating instantly all organic structures that amino acids form, as peptides, polypeptides, and nucleic acid.

Can't amino acids just change their structure to resist oxygen, and create different combinations of reactants within their side chains? No, and this is the problem, because in order to do so, in order to reshape amino acids, life has to die back down to ions and simple radicals, to the ionic form of life, losing all these precious specialized intelligences by the zillions, and it can never happen, because these intelligences themselves are life. Yet if life had not survived the oxygen event over two billion years ago, it would have died down to the ionic form of life, and it would have morphed again, starting back with different types of amino acids at its roots, different than its ordinary twenty-two proteinogenic amino acids. Yet the old life never died, and therefore the old proteinogenic amino acids are the only ones still used, along with all the old intelligences that they carry, even in the current environment of Earth, as they are always protected within cells and organisms. While if you search very closely throughout the lakes, rivers, seas, and oceans of the current Earth, you will find communities of molecules based on other types of proteins, capable to live life in the current aerobic environment of Earth. This is the new life on Earth, always capable to reach the niches of Earth, and always capable to interconnect continuously harmoniously with the old life of Earth that you know well, currently living life as organisms in the organic form of life, and as cells in the cellular form of life.

It seems that the cellular form of life was strong enough to survive the oxygen event by adapting into the new prokaryotes and eukaryotes, which is not entirely true, because it was the cellular form of life to have created the oxygen event in the first place, and therefore only that specific part of organic life had survived. Was it a hard loss for Life? It was a gain and a loss, because life is very diverse. Life favors strength and intelligence even over consistency and vastness. Strength, ingenuity, and development make life survive, and through them, it makes Life lower the entropy in the universe.

Life creates the environment in all its niches, along with all living beings meant to interconnect and subsist throughout the environment, along with all intelligences helping living beings do so, and along with all the necessary specialized cognitive abilities allowing intelligences to subsist and interconnect within any environment, even if they have to shift their entire form of life continuously, higher and higher. Until the Consensual Matrix throws a monkey wrench in all these, and now this is the actual life, consensually contorted, as you make it here on Earth throughout courts, cartels, districts, radical political parties, mobs, lodges, and jurisdictions, affecting all these little ions and proteins.

Many times, there is only one choice for life to develop, only in that specific manner, as though existence is not random but unique, regardless of the multitude of environmental conditions and constraints that you find everywhere. Furthermore, if your development and achievements are not as previously intended, Life discards you even if you are an entire species, world, or civilization, and others come in your place to do everything all over again, this time better. Until the Consensual Matrix throws another monkey wrench in all these, to take life for itself.

Many times, it seems that life has a predetermined purpose for you, it watches you continuously, and if you do not succeed throughout all your intelligent human needs and meanings exactly as intended, since your intelligences know well what you have to do, she can't wait to remove you and make an example out of your failure, inability, weakness, consensual servitude, and neglect. Since if you live your life consensually, addicted, diverted, and in ignorance of all these, you do not stand a chance. While many times, there are people who constrain you to become this way, throwing you under the bus of Life, which is a big bus, while this is an implicit form of genocide.

The chloroplasts from within our eukaryotes are actually cells, not organelles. They are regular prokaryotic cells, living beings of the third form of life, which is the cellular form of life, specialized in converting solar radiation into electrons, charge, and energy. The structure within the chloroplast specialized in this task is called thylakoid, and this resembles to a solar panel. Yet the thylakoid is simply a transparent membranous structure holding in place specialized proteins, along other auxiliary structures. Proteins capture electromagnetic radiation coming from the Sun directly, with their little helices, condensers, and other living electronic devices transforming it into charge, field, and electricity, while using it to strengthen up the field, or directing it to other structures to form ATP molecules, sugars, and amino acids.

The only problem is that proteins are measured in nanometers, while wavelengths of visible and ultraviolet light are measured in hundreds of nanometers. Proteins are several hundred times too small to see light or to stop light, let alone to harvest any energy carried by light. Yet visible light passes straight through many cellular components, since its wavelengths are larger, and this is why you have to add ink in order to see cells and larger cellular components with a microscope.

How do proteins manage to see and stop light, since they are hundreds of times smaller than its wavelengths? The thylakoid itself is larger than the wavelength of visible light, with many of its lamellae matching exactly the size of the most intense wavelength of light, the color green. However, lamellae and the entire thylakoid are only holding structures, but they do not capture light, they only hold proteins steady in place. While the proteins themselves capture the sunlight energy. Why do proteins have to be held in place to do their job?

More importantly, how do proteins do their job, being so small? Proteins use what we call in optics interference. When two or more waves interfere, they form what we call standing waves, places filled with condensed light in some places, and null places in others. It happens that these specific places of interference are smaller than the wavelengths of interfering waves, and depending on how you create the interference, you can make them as small as you want, even matching the size of a protein, or the size of one helix of a protein more precisely. While you have to place the protein with its helices exactly in the place of intense interference, and you have to hold them steadily there. This is why you use an entire hard transparent structure in order to hold in place all proteins harvesting sunlight through the process of interference. Later on, the organic form of life adapted this entire living process of photosynthesis in eyes, allowing the same specialized proteins to record images live, while they even used these specialized proteins to block sunlight in the skin, as a pigment everywhere throughout the skin, in order not to damage the organism.

There are many ways causing waves to interfere, including letting light pass through narrow fringes, and it happens that the same proteins also create the necessary fringes. These standing waves formed by interference are fixed in space, and therefore all proteins must be placed in a specific pattern, in order to match precisely these specific places, while you must hold them there very still. This is why you need a solid, transparent membrane to hold proteins in place while harvesting energy from light, and this is what the lamellae and the entire thylakoid do.

Cyanobacteria looks green, making all plants and algae look green, because chlorophyll and thylakoids are green, in order to capture the strongest wavelength of light, which happens to be green. You might assume that the color green is given by these specialized proteins harvesting the energy of light, which is not true, since proteins are hundreds of times smaller than the wavelength of light associated with the color green. It is only through the phenomenon of interference that proteins give the impression of green, as holograms from credit cards give the impression of forms and colors, since they use interference similarly.

We notice that living molecular communities must have a necessary basic structure to survive, along with intelligence storage and preservation in form of RNA or DNA, along with workers in form of proteins and enzymes, and replicators in form of ribosomes, which are also proteins. Along with a power source and all the necessary means to use it. Within cells, this power source is either the thylakoid harvesting energy from the Sun through photosynthesis, or a place in the cell where incoming chemical compounds are synthesized into energy, carbohydrates, and amino acids through chemosynthesis. These power sources define the type of the entire living community, its type of food chain, and its place within the food chain and within the rest of existential chains. Most of these power sources within cells use chemosynthesis, by harvesting energy and electrons from chemical reactions, the way it had been the case throughout most of our model until sunlight touched the surface of Earth. They use photosynthesis now, by harvesting solar energy. Animals feed either on chemosynthetic life or on photosynthetic life at the base of their food chain.

Life can be understood throughout interconnectivities as a continuous competition, or as a continuous harmonious cooperation. Living beings, if they are not on different levels of the same food chain, always cooperate, by sharing knowledge and resources. We are going to study a specific case of cooperative interconnectivity, which happened right within the oxygen event, over two billion years ago.

When cyanobacteria had first appeared to harvest solar energy, spreading the oceans and the surface of Earth, it took them some time to poison the environment with oxygen, probably hundreds or thousands of years. This is how prokaryotes and wall-less molecular communities managed to remain in existence, and even cooperated with cyanobacteria and other prokaryotes. Old communities of proteins provided the cyanobacteria with raw reactants, while cyanobacteria provided them with electrons, with ions, with amino acids, and probably even with ATP molecules. Specialized proteins from both sides appeared over time, making the transfer of energy and resources between communities and cells by the thousands. These proteins held their genes within specific RNA molecules, now growing in number, size, and priority. This happened at all levels of all food chains, to all living communities, chemotrophic, photosynthetic, or communities feeding directly on other living communities. In the meantime, more wall-less living communities developed cellular membranes, again, to keep within themselves a mild environment of low levels of oxygen to all members. This is how, many newly formed cells included directly entire other cells, cyanobacteria, other prokaryotes, and even other communities of molecules, only to keep cooperating with them, as powerhouses, reactors, and data archives, for the entire newly formed larger cell. These cells found within the larger newly formed cells are called currently the chloroplast and the mitochondrion, while the entire newly formed cell is called an eukaryotic cell. Currently, only eukaryotic cells form organisms, not prokaryotic cells, which could mean that the eukaryotic cells had been formed with the intention of forming organisms.

It is said that the chloroplast ant the mitochondrion live in symbiosis within the eukaryotic cell, which is not correct. This is not a symbiotic relationship, since the eukaryote is not exactly the initial living community. The eukaryotes are not hosts, since the chloroplast and the mitochondrion had been within the eukaryotic cell from the beginning of the eukaryotic cell. These are several initial cooperating communities living within one boundary, having one cellular membrane, forming a larger living community, simultaneously forming a unique living being of the fourth form of life, the eukaryotic form of life. Similarly, all cells of an organism do not live life in symbiosis with the entire organism, but they are the entire organism altogether, only in a higher form of life, in the organic form of life.

There are sometimes only one prokaryotic cell living within one eukaryotic cell, sometimes several, dozens, even thousands of prokaryotic cells living within only one eukaryotic cell. Note how, again, cellular membranes appeared around living communities only as a response to a significant change in the environment.

This is what I refer to as Life folding upon herself

whenever it forms a higher form of life, when living beings of a higher form of life are formed by a mixture of beings of various lower forms of life, only to maximize variety and therefore to diversify intelligence, making it more intuitive, more intelligent, and therefore more capable and more pertinent. This is the case with eukaryotes, which are living beings of the fourth form of life.

Eukaryotes live within organisms currently, forming living beings of the fifth form of life, the organic form of life. Before we start studying life at this fifth form of life, let us see why and how it had been possible for cells to gather into tissue, organs, and organisms. We are past the oxygen event now, and we pause the model around one and a half billion years ago.

Cells never live alone, but always alongside other cells, mostly their own kind. Bacteria in general, fill up the space everywhere where they live with their own kind, through sustained division, expanding in this manner by the trillions and more in little tight places overcrowding everything, until they finish all food and resources. These amalgamations, formed mostly by similar individuals, are not actually living communities, but cooperating communities. With the difference that the first ones are living beings holding unique intelligences through higher forms of life, and the second ones are simple gatherings.

All cooperating and competing communities, along with all individuals, the place that they inhabit, and their resources, form an ecosystem. Rational intelligent living beings do not adapt to their ecosystem anymore, but they change the ecosystem itself while they fulfill their needs, many times entirely. Even ecosystems of inferior living beings are changed by all inhabitants, implicitly and less significantly. This is how the environment is changed slightly at the interior of an amalgamation of cells. The cells living within the amalgamation benefit of a safer, non-changing environment, while the cells at the periphery of the amalgamation face the elements and the competition, they receive more food and resources, yet they are targeted by predators and accidents more often. The cells at the periphery of the microbial culture are more active, they receive more resources, they learn and adapt more, and they tend to act on behalf of all cells of the culture found at the interior. The interior cells obtain less resources, yet they can invest more time and energy on behalf of the entire community by becoming specialized, helping in this manner the cells at the periphery of the community. This happens not with eukaryotic cells that are already specialized within the organisms that they form, but with prokaryotic cells that do not live within organisms but only within cultures or communities of prokaryotes. While they still specialize similarly to the eukaryotic cells, only in a lesser manner, just as humans specialize within the human society, while they still look alike even though they perform different tasks within society.

Again, we see a real community forming with prokaryotes, even if there is no skin around the entire amalgamation of prokaryotic cells to define it permanently as an organism. In time, the cells at the periphery of the prokaryotic community divide and remain at the periphery, while interior cells divide and remain only at the interior. This is what develops specializations, while we even have one living relic from those times to tell the story, the water sponge.

Water sponges are not organisms but organs. They are simple gatherings of cells of the same kind living together, even under the same skin. What they do, they gather into large furnaces within which they pomp water with their flagella while filtrating the nutrients. Every cell receives nutrients in this manner, through a collective effort.

Why is the water sponge not an organism? Because you can disassemble the water sponge into individual cells and they keep on living, probably having to work harder to find food. Then later on, they can reassemble into the water sponge. The water sponge resembles more to a living individual organ than to an organism.

Forwarding the model now, we notice the formation of the DNA molecule within eukaryotes, which is a larger, very stable structure meant to hold prototypes of RNA molecules, the way each RNA holds prototypes of each type of protein. The DNA structure appears because there are more new types of specialized proteins, meant to perform all tasks related to the management, communication, interconnectivity, and synchronization of all cells in an organism. Cells are synchronized by hormones, which are mostly proteins.

What follows next in our model is normal taxonomy, with classes and species developing out of previous ones, according to the environment, and according to their abilities. Taxonomy states correctly all species of the organic form of life, only that old species do not develop into new species, but only specific groups of living beings from the old species develop together into new species, but not the entire old species. This means that the organic form of life is not divided into species that are divided into living beings, but the organic form of life is formed of living beings directly, while these develop, and through them, the entire species develops.

Note that it is not so easy to develop from on species to another or from one class to another. Development is the summation of all new solutions and the effects that they have on the entire species. In order for any solution to count throughout development, this solution must be significant enough to be shared with all members of the species, and therefore to be added to the genetic sequence within the DNA or RNA. When new solutions are only partially shared within the species, even if they are very important and they make it in the genetic code of a part of the species, they are erased throughout future generations, being identified as genetic errors. This is how it is very hard to share new solutions within an entire species and to develop into a new species, mostly when the species is very large, spanning large areas of the world, which is the case not only with humans, but with almost all species around us. This is how, very successful species covering the globe end up learning and adapting less, while stagnating. While when they are challenged by major events, they tend to perform less than younger, emerging species, and they go extinct.

We have two subjects to follow throughout taxonomy and therefore throughout time from the raw field to human beings. We must follow both the physical body and the intelligences within. Because when you follow the physical body, as we have done in this chapter with molecules, mostly proteins, and then with entire cells when these developed on Earth changing the entire molecular form of life into the cellular form of life, you notice how entire species of life develop, as the multitude of types of cells spanning the world ocean after the oxygen event. However, when you study molecular intelligences, as protein intelligences, RNA intelligences, and later on entire cellular intelligences, you notice how not exactly the actual species develop, but these specific intelligences do, continuously, within a continuously shifting environment. Which means that not exactly species evolve, but all or some living beings within these species develop, simultaneously, from the dawn of life, in only one generation. This difference is not too noticeable right now at the level of the cellular form of life, yet later on, within the organic form of life, you cannot consider anymore that species evolve, but only that individual living beings develop, all the way from the dawn of life to their own generation within their own species of life.

While we also notice this entire, continuous, meticulous development of life taking place in order to match as much as possible the particular niches of the environment making it possible, with the greatest efficiency. Life herself takes you out if you remain incapable to reach and match your own niches at maximum efficiency. This is why you find all these molecules, communities of molecules, cells, communities of cells, organisms, and communities of organisms striving to reach the available niches of Earth in the most efficient manner.

When other intelligences find better solutions to reach niches in an even more efficient manner, then all homologue intelligences of all living beings of all species of life learn to do the same in a cooperating manner, increasing their efficiency together. We also notice a cooperative behavior and a competitive one among all living beings, with the cooperative behavior making all species possible, and with the competitive behavior taking place at times throughout extreme unfavorable environments.

Then the environment changes again, all niches tilt, with all intelligences striving to find even more efficient manners to reach them continuously, and this adds to the continuous shifting of life to newer species, newer classes, and newer forms of life. There is more to study here than the trivial answer that capable species live while the incapable ones go extinct, as the theory of survival of the fittest states.

There is even more to consider when you study life, living beings, and development through its intelligence correspondence, because intelligences do not die with the death of their physical bodies, because intelligences are not individuals but they are systems of intelligences in themselves, resembling more to communities and civilizations than to living beings. Because all intelligences are systems of intelligences in themselves, to the point where you cannot find elemental intelligences anywhere in the world, but only systems of intelligences, making all intelligences communities or civilizations of living beings, but not individual living beings. While in contrast, you can always distinguish life in its distinct cells or distinct living beings when you study the physical bodies.

Physical bodes replicate when they are molecules, they divide when they are cells, or they reproduce as organisms, while intelligences are always systems of intelligences, and therefore intelligences always reproduce through colonization. Which makes intelligences immortal, only colonizing new living beings at the moment of conception, as new living molecules, new living cells, and then new living organisms.

Intelligences are immortal, and in this manner, every living being cannot be born, appear, or be built from scratch with each cellular replication or with each organic generation as it seems from a physical perspective. It is not too hard to see how the same cellular intelligence ends up in both daughter cells after the cellular division, since the parent cellular intelligence simply splits in two as a civilization, into the two new intelligence civilizations of the two daughter cells, and then the two cellular intelligence civilizations grow shortly to their parent size.

Yet the same happens with each newly born organic living being at conception, since the new organism is not constructed from scratch with each generation as you might assume from a physical perspective, but it is a normal intelligence colonization as the newly born organism, since all intelligences expand in this colonizing manner at the moment of conception. Which means that all newly born organisms have to develop themselves from the dawn of life with each generation, during the short time of gestation, exactly in the manner that the entire life had developed from the dawn of life, from the ionic form of life, since this is the only manner that intelligences of the ionic form of life know how to grow and develop while forming molecules, cells, organisms, and entire ecosystems.

You can study all fetuses throughout gestation, to notice how they transition gradually through all species that these intelligences were before, one after another. Since this is how you can find living tyrannosauruses by the end of gestation of each normal chicken egg, with reptilian claws, reptilian teeth, reptilian brains, and reptilian wings included. While the current esoteric science is capable to stop the chicken gestation right there, in order to bring to life exactly these flying reptiles of the past. You also find insects, reptiles, and amphibians at various moments of gestation of the human fetus, since all humans and human intelligences were these in the past ages of Earth.

Because all intelligences are systems of intelligences, resembling communities and civilizations of living beings, and this is how they develop. While every time they give birth to new living beings, they do not build these from scratch in their own current image, but they have to develop themselves through each one of them as they did ever since the dawn of life, ever since the ionic form of life.

Which means that species never evolve, since this is only the appearance, but all living beings develop individually from the dawn of life during gestation, to become when they are born what their species is.

Cataclysms are certainly unfortunate, yet they are opportunities for life to develop. As long as the environment shifts its multitude of niches as it pleases, life follows closely with its highest efficiency possible, regardless if it wants it or not. Searching throughout history, it is during and immediately after cataclysms that you find species and classes of species developing, and even entire forms of life. This is how one extreme cataclysm on Earth had caused the sudden development of reptiles into both birds and mammals, simultaneously, almost three hundred millions years ago.

As just seen, intelligences do not die from one form of life to another, from one species to another, or from one generation to another, but they simply carry on, even indefinitely, in their specialized manner. Furthermore, as seen throughout this book, intelligences, along with the physical bodies holding them, cannot change themselves too easily, while they cannot exactly construct themselves from scratch, but they have to develop continuously exactly as they know how to do.

This is how the new generations of the organic life are not exactly built altogether by using the information stored in the DNA, as science implies, because there is no information in the DNA, but only intelligences, specialized intelligences, or prototype specialized intelligences, while these always do exactly what they always know how to do.

Which means that, with each generation, you have to develop all these intelligences from the very beginning in order to give birth to organic living beings with each new generation, because these are the intelligences of the ionic form of life, and you cannot have them otherwise. While the intelligences of the molecular form of life, cellular form of life, and organic form of life are not exactly intelligences, but they are entire systems of intelligences or societies of intelligences, formed only by the ionic intelligences of the ionic form of life by the zillions. You have to develop all these intelligences from the dawn of life altogether, in every living being throughout gestation, as developed as they can be, which is an achievement. Which means that species do not evolve into new species, because all intelligencers are part of living beings and not of entire species.

As seen, each living being develops during gestation from the dawn of life to its current form of life, class, species, and level of development. Amphibians are even born as fishes, while throughout gestation, they develop throughout the entire invertebrate life. It is the same with insects, since they are born as worms. While with the human fetus resembling fishes, birds, reptiles, and many types of mammals throughout gestation, now this is how the entire organic life actually lives life.

Because the DNA is not the abstract blueprint of life, since you cannot store directly abstract information within the nucleic acid about life from one generation to another and from one species to another, but you can store only living prototype intelligences within the genes of the DNA and RNA, and these know exactly how to arrange all cells within the organism throughout its growth in order to bring it to its normal level of development. These are the developmental intelligences. Furthermore, the same intelligences making you grow up exactly in the image of your parents and entire species, send you now the specific sets of developmental needs meant to render you an intelligent human being, developed at the third intelligent level, along with all the intelligent human needs necessary to determine you to build around yourself the intelligent human environment, including the intelligent human society, while these developmental needs are genuinely alive.

Therefore, species never evolve from species, but living beings develop from one species to another, together, whenever the environmental conditions demand and allow. This is how life is not exactly lived at the level of species and entire higher forms of life, but it is lived at the level of the ionic form of life, by all intelligences of the ionic form of life, interconnected continuously at the molecular level, cellular level, and organic level, while forming together the molecular form of life, cellular form of life, and organic form of life. Life is lived at the level of the ionic intelligences within, regardless if the ionic intelligences interconnect to live life in higher forms and classes of life. While the physical body is only the correspondent physical image that all intelligences have from an objective, physical perspective.

The current science still follows the theory of evolution, which is rather primitive, having only one statement, that species evolve. Which is erroneous, because species do not evolve, but individual living beings do, in one lifetime, mostly during gestation, starting with the dawn of life. While throughout history, not the species evolve, but large groups of living beings within the old species do, into a new species, but only if the environment demands, with the rest of the living beings remaining in the old species.

You cannot produce living cells from scratch, but you have to divide them out of old cells or stem cells. Cells replicate or duplicate their members first, and then they divide. It is the same with living beings of the fourth form of life, eukaryotes, since they have to start from the bottom, duplicate all molecules composing them, divide all composing prokaryotes, and then divide themselves. While with the reproduction of each organism, you must replicate, divide, and reproduce each living component and intelligence counting in zillions, starting with the dawn of life. You do everything as it had always been done starting with the dawn of life, since this is all that your intelligences know how to do, but you do so during gestation and little after, because insects are born as worms, and frogs as fishes.

Considering an organism a genuine living being of the fifth form of life is as considering an ant colony to be a genuine living being of the sixth form of life, which is only partially true. As you study an ant colony closely, you see it formed of the same types of individuals, all carrying the same genetic code. That is a living family, made possible by a unique individual, the queen, reproduced now thousands of times or tens of thousands of times, since all her ants are also her children. It is the same with organisms at their fifth form of life, since organisms are formed of the same egg cell, divided a zillion times throughout gestation, and then divided even more during the growth of the organism after birth and throughout life in general. Because you cannot reproduce organisms, but you replicate the ions and small molecules of the ionic form of life within the sex cells of the entire organism, and then you must develop these very fast throughout gestation from one form of life to another and then from one species to another in the organic form of life. Yet this is also the case with human societies and human civilizations, because you cannot reproduce these, but you give birth to a multitude of human beings, and then you restart the human civilization from scratch by helping them learn and interconnect exactly as their developmental intelligences demand.

The same developmental intelligences that determine entire cells to grow after division to resemble the parent cell can arrive to specialize in the growth of the entire organism, developing it throughout gestation and growth just as well, formed of similar specialized physical structures, allowing it to fulfill the same specialized inner and outer needs throughout life. Then the same developmental intelligences can end up specialized in the development of the entire society or entire ant colony, structuring this sixth form of life through similar specializations, which allow the entire society, colony, world, or civilization to subsist, through the fulfilment of the same specialized needs, by similarly specialized living beings.

This is the case if this sixth level form of life is allowed to behave and develop freely, naturally, and in a living manner. Because as seen, the higher you are allowed to ascend in forms of life, the closer you get to Life at her last supreme form of life, completing the circle of life, and becoming a highly developed living being yourself.

It is the same life, and these are the same intelligences present in all forms of life, regardless if they live life at the first form of life, at the third, the fifth, the sixth, up to Life herself, because this is the only manner to develop and subsist, by going throughout the entire life development in one generation, in one lifetime, if this is the only manner in which the wider world allows it, as it is the case with this specific organic form of life here on Earth. Homemade appliances, as always seen, but now at a supreme level. While this is a glimpse of the tenth supreme level, of Life herself.

This tells something, that primal intelligences never die and never reproduce, but only manage to transcend their consciousness from one generation to another, from one species to another, and from one form of life to another. While this is the only way that this consciousness knows to develop at each level, as an individual, species, and form of life, through the actual development of life itself, because this primal consciousness is life itself, only truncated or projected at its own level.

As seen above, when you study the organism, you find it similar to societies, communities, civilizations, and colonies of similar individuals, since all cells of an organism have a similar origin, the fertilized egg, and therefore they have a similar genetic code. Organisms are not a genuine community, but a genuine family, since organisms grow out of one single cell, just the way one queen gives birth to her entire colony. If you want to duplicate an ant colony, you must start with the queen, let her give birth to another queen, and the new queen will form slowly the new colony, which is never put together part by part and ant by ant as from a blueprint.

It becomes even more complex, since as seen, throughout gestation, the newly growing organism has to follow the same developmental intelligences within the DNA that its entire species had followed since the dawn of life, up to where it is currently throughout gestation and growth. This is why organisms must be recreated from the origin, not only from the last generation, but from the very origin of life that took place billions of years ago, in one lifetime. This is the case because you cannot rebuild or reconstruct primal intelligences with the birth of each living being, because primal intelligences never die.

With all developmental intelligences following this same

developmental pathway throughout gestation as it had been the case throughout the development of the entire life and organic life since the dawn of life, not only because this is the only thing that intelligences know how to do, but because all these developmental intelligences were right there themselves and right then throughout the development of the entire life since the dawn of life, making all these bodily changes and cognitive changes possible right then, with each change in the environment demanding them persistently. Therefore, this is what they do during gestation while developing the fetus itself, since this is the only thing that they know how to do.

Why does this happen? Why exactly cells stopped adapting to the environment? Because they have very developed intelligences, capable to reshape the environment in every way necessary to fulfill their needs, while they already carry their own environment with them entirely, in their portable bags, called cells. This is the entire developmental process that science calls evolution, the continuous developmental reshaping of the environment of these specific primal intelligences, because they, along with their own physical bodies, are already sufficiently developed. Whenever they need to adapt to the environment, they do just as humans do, they adapt the environment to their needs instead. This is how they end up adapting all higher forms of life found around them, their molecules, cells, organisms, communities, societies, species, civilizations, ecosystems, worlds, and realities, but not themselves, since the laws of physics and chemistry stop them.

Why do intelligences have the need to adapt? Because now, as entire organisms, these cells and cellular components have to keep the continuously unfavorable environment further and further away from them, through the shape, size, strength, and ingenuity of the entire organism, family, genetic line, community, nation, and society. While coincidentally, many of these same cellular intelligences manage to develop, interconnect, and transcend one higher form of life after another, to reach as high as their own natural needs take them.

This is what life is, and this is what larger and larger forms of life as organisms, species, communities, nations, societies, and entire civilizations are used for, to keep the increasingly unfavorable environment away from the living ions, molecules, and cells, and more precisely, away from the ionic intelligences of the ionic form of life living life right on top of the electromagnetic field. This is how life manages to exist at all levels and in all forms of life simultaneously, because the ionic intelligences manage to form entire cognitive systems that span the larger and larger forms of life. These cognitive systems spanning entire forms of life can be considered individual intelligences, while the forms of life that they span are also considered individual living beings.

This is the case with you, the conscious intelligence, since you are a cognitive system in yourself, filled with inner intelligences that are cognitive systems in themselves. While your organism is a form of life in itself, filled with trillions of cells that are forms of life in themselves, cells filled with cellular components, molecules that are forms of life in themselves, and there might be life at atomic level, nuclear level, down to the raw field itself, life living in the raw field, and from there reaching out of this world to span the entire wider world, one higher reality after another. Because life is interconnected, life is everything, everywhere, and everyone.

Why do we look the way we do? There is not a great difference in the shape of an organism from one species to another. This is why science is impressed by the genetic similarity between classes and species. The cell, the initial egg of life, divides itself in specific ways and forms to match the environment, to become the fully-grown adult organism, forms that take little genetic data to define. There is not much development taking place from one species to another, only that cells are arranged in a slightly different manner.

This is why you do not find specific abstract blueprints and schematics within the DNA on how to build an organism from scratch throughout gestation, the way biology and genetics teach, but you find genuine developmental intelligences capable to handle the entire development of the new living being starting with the initial cell, the egg of life. These prototype developmental intelligences from the DNA are the same specialized intelligences to have developed life in that particular manner in that particular moment of time in the past when it happened, and now this is the only thing that these developmental intelligences know what to do. While these intelligences already have the necessary abilities to make the fetus develop successively, into the necessary structure and shape forming exactly the organism of that specific species at adulthood.

Each procedure of growth for each organism is not elaborately rewritten with each evolution from one species to another, since that would involve very advanced knowledge and abstract reasoning, absent throughout the development of organic life. The knowledge regarding the transition from one species to another is simply added raw on top of the knowledge regarding older development from the original living community of proteins or from the cellular development to organisms, one on top of another, just by allowing specific developmental intelligences to grow the entire organism one form after another, exactly as they have done throughout billions of years and more. Because these are the same developmental intelligences growing the entire organism now, as they had managed to develop life altogether from one species to another and from one form of life to another, billions of years before. This knowledge and intelligence had not actually been specifically created or modified, to be used during gestation for the growth of individual organisms, but these are exactly the same knowledge and developmental intelligences used by every species to live, behave, adapt, and develop one form of life after another, starting way back with some ion in the field, then with a living molecule, or living community of molecules, and ending up with the current species and the current generation, in a flawless manner, without degenerations. This is the baggage of solutions for old problems shared among living beings, and transmitted accurately from one generation to another through the exact specialized intelligences developing in the first place, now stored as prototype intelligences within the nucleic acid, because this is what the DNA or RNA hold.

The DNA and the RNA do not even hold data, blueprints, tables, and plans, but they hold only what type of protein to be built or to be born for what type of task, need, and problem encountered, along with its protein intelligence intact and alive in it, how many proteins to make in one batch, how many batches to make and exactly how to assemble them in order to perform that specific task, in order to solve that type of problem. Because this is how solutions to old problems are saved into the nucleic acid, in form of prototype protein intelligences that already know how to solve these problems and how to fulfill these specialized needs, including developmental needs, depending on specialization.

Now you can understand genes, RNA, DNA, cells, organisms, origins, creators, natural laws, supreme laws, Life, learning, development, reasoning, logic, intuition, intelligence, societies, needs, meanings, feelings, behavior, worlds, realities, field, species, intelligences, consciousness, existence, interconnectivity, life, evolution, and living beings, since these are correlated with the specific needs and abilities of all your intelligences, all being included in this specific manner within the abilities and knowledge of all intelligences of your cognitive system, all acting freely right now throughout your organism while performing their tasks or only waiting, stored as prototype intelligences within the genes of your DNA, for further use. The proteins themselves stored in the genes of the DNA and RNA know how to solve all major previous problems encountered by cells and organisms in the environment, and how to perform all specialized tasks within cells and organisms. They know how to do everything because their specific intelligences are the ones knowing exactly how to do so. The protein intelligences are the specialized intelligences of the cell, or only colonies and colonists of these.

These intelligences are stored alive within the DNA for a later use, since they know how to form again the entire cell after division, along with the entire organism after conception, along with the entire community from scratch, and along with the entire herd, pack, society, civilization, world, reality, up to Life herself. They know everything, because they have been there before, and they know well what to do. They can do everything by performing their own specialized tasks, exactly when they receive their needs. While when they cannot do anything themselves, they know how to send needs and feelings to other intelligences by their specialization to help, to cooperate with them while they help, and even to constrain them to help if necessary.

They send similar needs to you the conscious intelligence, to do your part and to fulfill your needs in the outside world. While you fulfill them, since this is what you do throughout life, you fulfill needs, tasks, and meanings, natural and artificial, everything related to the outside world. You know exactly how to do everything, and when you cannot, you just have to use your logic, intuition, and intelligence throughout mental models to figure out everything. While you have the entire cognitive system to help you throughout your new conscious cognitive tasks, just as you help the entire cognitive system, organism, society, and the entire ecosystem through the multitude of needs and tasks that you fulfill on their behalf in the outside world, because intelligences and prototype intelligences stored alive within genes know what they have to do, since they did everything before.

While as you already notice, from your own perspective, it makes no difference if your own developmental intelligences send you specific needs to better your diet, to better your mean of transportation, or to better your behavior in society, because you are always determined to fulfill all needs and meanings as they are and as they come, enhancing your survival, subsistence, learning, and development.

You have to eat when you receive your need to eat, and this is why you eat with your main meals, while feeding your entire family. You have to learn when you receive your need to learn and develop, and this is why you read these books. You have to improve your home every time you change and rearrange your furniture, making your living conditions better, because this is your family developmental need that you receive and have to fulfill. Your own developmental intelligences send you all your needs throughout life to help you survive, subsist, learn, and develop continuously.

While you do everything on your behalf, developing yourself continuously, and on behalf of your entire family, tending to every member of your family harmoniously, exactly as they need, and more importantly, exactly as your own developmental intelligences send you your needs and meanings to tend to your loved ones, to your family and friends, and even to members of your community, city, nation, and the entire world. While the same developmental intelligences send you similar developmental needs of very high classes demanding that you tend to your entire society and the entire world in a living harmonious manner exactly as you do at home, as though every living human being is a family member, is a loved one, and you always feel it. This is commonly called care, as you always care, as everybody else.

Since you always care when you see all loss, suffering, death, and destruction in the news, or when you see people starving throughout the poor nations. Because all your developmental intelligences, just as they develop you flawlessly from simple cellular components to the living human being that you are in only a matter of months and years, just as they know what needs and meanings to send you to make you capable to survive, subsist, learn, and develop alongside your entire family, community, and entire world, your developmental intelligences send you very specific developmental needs meant for you to develop your entire home, family, community, society, civilization, world, reality, and the entire human environment, exactly as they had these throughout their own lives, cognition, experience, lives, and development.

They never leave you alone until you fulfill all your needs, to tend to yourself, your family, community, and entire world, and until you make these as prosperous and as developed as they had them throughout the best times of the human history, the famous human golden ages. Since as stated, life strives to reach and cover all niches offered by the current environment as efficiently as possible and for as long as possible, with the golden human ages of Earth being the most prosperous, the most efficient, the most harmonious, and the most developed.

The golden human ages might have been here on Earth in the past, or they might have been in other worlds and realities throughout the past existence of the organic life, and now your own developmental intelligences will never leave you alone until you make them possible again, alongside all living human beings of the world, since everybody receives developmental needs of higher levels and higher classes.

Yet with the Consensual Matrix working overtime to stop you, erasing systematically the entire living intelligent human environment continuously, you never reach the golden human ages, but you remain continuously in the dark ages. All living human beings remain determined intrinsically to instate the intelligent human environment, to maintain it, and to improve it in the entire world continuously throughout their life, since they have all intelligent developmental needs in them, yet by taking drugs, they interfere with the entire human cognition while ruining everything, never reaching the actual intelligent human level, along with the comprehensive intelligent human golden age.

This is why the current consensual science teaches you that successful species survive but the unsuccessful ones die away and now this is how life develops, that you think randomly with your brain, that you must always take your medicine, that this world appeared spontaneously randomly, that you must always have only one spouse or no family at all, or that higher abilities are science fiction, and now this is the human meaning in life and in the word, so you can always serve tyrants. With everyone else feeding you drugs and ideologies continuously because they are just as undeveloped, while you serve tyrants diligently and while you erase some more the intelligent human environment, standing in the way of everyone who is slightly more developed than you and seeks to fulfil their intelligent human needs.

Which is not your case, since if you have made it all the way to the end of this book, it means that you are developed enough to reason intelligently alongside it. Be thankful that you still have the means and awareness to develop, and that you are still capable to find meaningful knowledge through all these books, for as long as you are able to find them, because as you look around, most of the world remains indoctrinated and under servitude continuously, while standing in your way.

This is why you care for the world, this is why you suffer when there is failure and loss in the world, and this is why you do everything to make a better world through the fulfillment of all your needs and meanings of higher level and higher classes, since this is the better world that your own developmental intelligences want you do have and achieve, as they had it before even continuously throughout the entire human life. As it is the case with all living beings and intelligences of all types and forms of life, all striving continuously throughout life to fulfill Life herself.

While from their own perspective, life is lived entirely at their own level and on their own behalf, exactly as it comes one need after another, regardless of what form of life they are, what species and individuals they are, what environment they form, and what specific age of Earth they go through. While these are actually the primal intelligences of each living being managing to live endlessly from one generation to another throughout genetic lines and throughout entire species, species after species, and form of life after form of life, ever since the dawn of life. You are one of them, the conscious intelligence, specialized in the coordination of the entire organism in the outside world, as it fulfills its needs, or as you fulfill your needs, since it is the same.

How many intelligences are there in the organic form of life and everywhere else here and throughout the wider world forming Intelligence, Life, the Field, Existence, and the entire world? What exactly do they do? How intelligent are they? How do they interconnect? What do they want? How exactly do they look like? Let us see.

The End

This book series continues with the next book, "The Hierarchy of Intelligences." Here is a short synopsis:

You are a conscious intelligent living being, yet you are more than an individual being, since you are an entire living compound made of distinct living beings, all forming you, while all contributing to your unique life and existence. Because you are your trillions of living cells, composing your body one after another, while with each cell, you are a zillion subcellular components forming you one molecule at a time and one cell near another, designing or creating your life one fulfilment at a time and one moment of existence after another, while animating you from within, from the smallest subcellular component.

This is who you are at the physical objective level, while all living beings generate, sustain, and coordinate continuously a multitude of intelligent modulations throughout the physical brain and throughout the entire body while forming your entire cognition, living intelligent oscillations associated to your mind, consciousness, awareness, and intelligence.

From the physical perspective of your body, these are simple encoded oscillations within your neurons and therefore within your entire mind, and everybody has them, capable to sustain entire inner cognitive instances forming together your cognitive system, your entire mind, in all its thoughts, awareness, feelings, and reactions. Yet from an inner cognitive perspective, all these physical oscillations become alive, making everything possible throughout your mind, organism, life, and entire world, in a living intelligent manner, since these are your intelligences.

All your intelligences are aware, conscious, and certainly

intelligent, while through their own consciousness and intelligence, they give you your entire mind, in all its cognitive abilities, becoming everything that you truly are. More precisely, your intelligences form your mind altogether, since the human mind is their living world, while your intelligences perform the entire cognitive activity within, for you and for the entire organism. Their continuous existence is your life, while their purpose becomes your meaning, achievement, learning, and subsistence.

Among all your intelligences, you are the conscious intelligence yourself, specialized in the coordination of the entire organism in the outside world while fulfilling its needs, or while fulfilling your needs, since it is the same. Because your intelligences have their own specialized tasks and needs, together forming not only your mind, but your entire life, in its entire existence. Yet these are only your own intelligences, the human intelligences, while throughout this book, we focus on all intelligences of the wider world.

This book studies all intelligences from living, cognitive, interconnective, and existential perspectives, throughout a comprehensive classification of everything alive and intelligent, while helping you understand yourself, life, intelligence, consciousness, the world, and your conscious place, meaning, and fulfillment in life and in the world.

ABOUT THE AUTHOR

Valentin Leonard Matcas, M.Ed., is a researcher, physicist, mathematician, educator, and an author of nonfiction and fiction books, including the entire "Human" book series. Valentin Leonard Matcas wrote the "Human" book series in the following order: "The Human Needs", "The Human Addictions," "The Hierarchy of Needs," "Stay in Shape, Lead a Healthy Life," "The Human Origins," "The Human Society," "The Human Conspiracy," "The Human Mind," "The Human Reality," "Astral Planes and Your Other Realities," "Life," "The Hierarchy of Intelligences," "The Human Intelligences," "The Human Thoughts," "Mental Models and Successful Ideas," "The Human Attitudes," "The Human Stereotypes," "The Human Ideology," "Modes of Life," "The Human Development," "Patterns of Development," "The Human Lifestyle," "Heal Yourself," "The Human Civilization," "The Human Religion and Spirituality," "The Human Rights," "Higher Laws," "Natural Laws of the Universe," "Existence," "The Human Condition", "Lifelines of Causality," "The Human Behavior," "Flat Earth," "The Human Environment," "The Human Meaning," "The Human Reasoning," "The Human Interconnectivity," "The Consensual Matrix," "The Matrix of Life," and "The Human Knowledge."
Valentin Leonard Matcas writes about terrestrial and alien civilizations, about life in the universe, the way it develops and intertwines across galaxies, about powerful beings as they control and reshape the universe, and about normal living human beings from Earth caught in this beautiful, wider, outstanding interconnectivity. Valentin Leonard Matcas creates a living, warmer universe in his books, teaming with life and vibrancy, on all levels of existence. Valentin Leonard Matcas also wrote "The Storyteller" book series, including "The Storyteller," "Starship Colonial," and "Unlimited," and "The Culling" book series, including "The Culling," "The Dream of the Dead," and "The Last Man on Earth."

Life

When he does not work on his books, Valentin Leonard Matcas enjoys researching, hiking, swimming, kayaking, skiing, snowboarding, biking, reading, listening to music, and playing strategy videogames. You may discover all his books, videos, and articles.

www.ingramcontent.com/pod-product-compliance
Lightning Source LLC
Chambersburg PA
CBHW020625220526
45464CB00001B/21